D0479245

THE

PRACTICAL BOTANIST

Rick Imes

All pictures supplied by
Photo/Nats

A FIRESIDE BOOK
Published by Simon & Schuster Inc.
New York London Toronto Sydney Tokyo Singapore

A QUARTO BOOK

Simon and Schuster/Fireside
Simon & Schuster Building
Rockefeller Center
1230 Avenue of the Americas
New York, New York 10020

Designed and produced by Quarto Publishing plc,
The Old Brewery, 6 Blundell Street, London N7 9BH

Senior Editor Sally MacEachern
Editor Sue Condor
Designer Anne Fisher
Illustrators Valerie Hill, Sally Launder, Kevin Maddison, Janos Marffy, Ann Savage
Index Hazel Bell
Assistant Art Director Chloë Alexander

Art Director Moira Clinch
Publishing Director Janet Slingsby

Typeset by Ampersand Typesetting Ltd
Manufactured in Hong Kong by
Regent Publishing Services Ltd
Printed by Leefung-Asco Printers Ltd, Hong Kong

1 3 5 7 9 10 8 6 4 2
1 3 5 7 9 10 8 6 4 2 Pbk.

Library of Congress Catalog Card Number: 90-45450

ISBN 0-671-69306-9
0-671-69305-0 Pbk.

CONTENTS

AGRICULTURAL LANDS

GRASSLANDS

WOODLANDS

WETLANDS

SEASHORES

DESERTS

HIGH COUNTRY

BOTANY AND YOU

FOREWORD

It is important to understand plants, in part because they make all other life on earth possible. Green plants are the only known organisms able to produce their own food internally through photosynthesis. This enables them to feed not only themselves, but all the animals which feed upon them, and predators that consume the plant-eaters. Can you think of any creature which does not directly or indirectly owe its existence to green plants? Trace any food chain back far enough, and you will always arrive at the common provider, green plants.

In addition, virtually all the oxygen in our atmosphere is produced by green plants as a by-product of photosynthesis. Several billion years ago, the atmosphere of our young planet was composed of methane, ammonia, water, and hydrogen gases. Driven by the sun's energy, storms raged across the surface of this barren world, depositing rain which accumulated over the millenia to form oceans. The lightning which accompanied these storms also sparked changes in the atmospheric molecules, producing larger and more complex organic molecules that eventually fell with the rain into the seas. It probably was in this primordial soup that one day, quite accidentally, a molecule arose that was able to make crude copies of itself out of the surrounding molecules. The copies continued to replicate. Those that were more efficient produced more copies, and it was in these dense masses of self-replicating organic molecules that the first single-celled alga originated and began to photosynthesize, releasing oxygen in the process. As the numbers and diversity of plants grew, the oxygen content of the atmosphere increased, stabilizing as the world became cloaked in greenery. Should a large portion of the earth's vegetation suddenly be lost due to some catastrophe, scientists estimate that, even if an alternative energy source were miraculously found to feed the world's animals, they would all suffocate within eleven years without green plants to replenish their oxygen supply. You can see from this that plants are indeed critical to the survival of all life on earth.

Right A food web illustrates the interconnectedness of all life forms, and at its heart are green plants, providing food and oxygen. Like a spider's web, a food web is strong when intact, yet easily disturbed, each disrupted connection weakening the web until it collapses. We must recognize the value of all species and the hidden relationships between them.

Left Plants do much more than ornament spectacular scenes, such as this view of Laryn Lake in Banff National Park, Alberta; they are essential to all life.

carnivorous consumers
herbivorous consumers
photosynthetic organisms
(producers)
birds
birds
insects
mammals
mammals
crustaceans
insects
algae
molluscs
amphibians
decomposers
reptiles

THE BASICS OF BOTANY

WHAT ARE PLANTS?

What differentiates plants from other organisms? Early biologists classified all organisms into two categories, the plant kingdom and the animal kingdom. This was based on relatively simple criteria: animals moved about and plants did not, plants produced their own food while animals ate other organisms, and plant cells possessed rigid cell walls lacking in animal cells. These distinctions worked well for the more complex, multi-celled organisms, but as more and more simple life forms were discovered and investigated, the two-kingdom system broke down. For instance, there are more than 300 species of single-celled, fresh-water organisms called euglenoids that are able to swim by means of flagella, or whip-like tails, and consume small food particles. These traits led them to be classified as animals by many taxonomists. However, many euglenoids also contain chloroplasts which enable them to carry on photosynthesis, suggesting that they belong to the plant kingdom.

As exceptions to the two-kingdom system accumulated, the need for more categories became apparent, but exactly where to make the divisions was unclear. Most one-celled organisms constitute a gray area in which it is impossible to draw distinctions without creating as many problems as you solve. To compound this dilemma, many unicellular organisms have left no fossil record through which we could trace their ancestry and clarify relationships. Today, more than six different classification systems exist, recognizing up to five kingdoms including *Monera* (bacteria and blue-green algae), *Protista* (protozoans, algae, sponges), *Fungi, Plantae* (bryophytes and vascular plants) and *Animalia* (multicellular animals).

Above Though fungi are now categorized in their own kingdom rather than the plant kingdom, they possess a number of plant-like characteristics.

Left The lush vegetation surrounding this Vermont farm may incorporate hundreds of species. Each is unique, yet all share the similar characteristics of members of the plant kingdom.

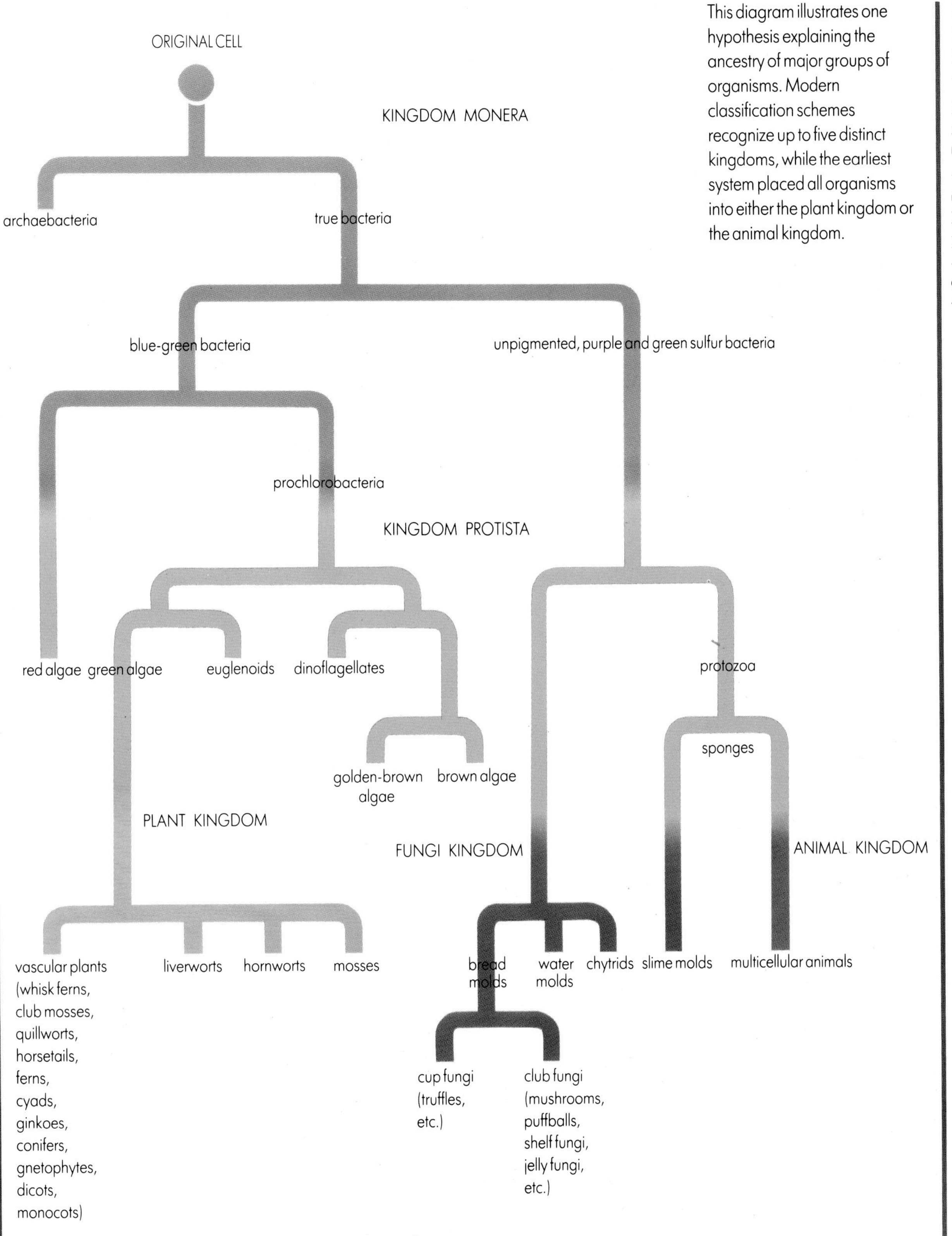

This diagram illustrates one hypothesis explaining the ancestry of major groups of organisms. Modern classification schemes recognize up to five distinct kingdoms, while the earliest system placed all organisms into either the plant kingdom or the animal kingdom.

PARTS OF PLANTS

There is considerable variation in plant anatomy throughout the plant kingdom. As a practical botanist, you will be concerned primarily with the largest and most obvious group, the flowering plants, and so it is there that we begin our study. Even among flowering plants, anatomy varies significantly, but they generally possess four major parts: roots, stem, leaves, and flower(s). Each part has different primary functions.

Roots

The root system anchors the plant and extracts water and essential minerals from the soil, transporting them upward into the stem. There are three major types of roots. Most *dicots* (dicotyledons, plants with two *cotyledons,* or "seed leaves") have taproots, while most *monocots* (monocotyledons, plants with only one "seed leaf") have fibrous roots. *Taproots* are large and fleshy, and often reach deep into the ground to obtain their needs. Taproots also store food, providing a distinct advantage to perennial plants starting into growth following dormancy. *Fibrous roots* depend upon branching filaments growing close to the soil's surface to collect precipitation before it sinks deeper into the ground. Many plants combine both types of root systems. *Adventitious roots,* the third type, are common to both monocots and dicots. These roots develop in certain plants, such as ivies, to facilitate climbing, and in plants with modified underground stems, such as corms, rhizomes and bulbs.

Stems

Stems support leaves and flowers, positioning them in the best locations to accomplish their respective tasks. Stems and roots also house a vascular system of xylem and phloem, tube-like structures which conduct water and nutrients to various parts of the plant. *Xylem* transports water and dissolved substances upward in the plant. *Phloem* can move material both up and down, and primarily transports newly-synthesized organic molecules, such as amino acids and carbohydrates, from the leaves to the roots and stem for storage, or to growing parts of the plant for immediate use.

Stems may be simple or branched, upright, or creeping. Many modifications to the stem enable

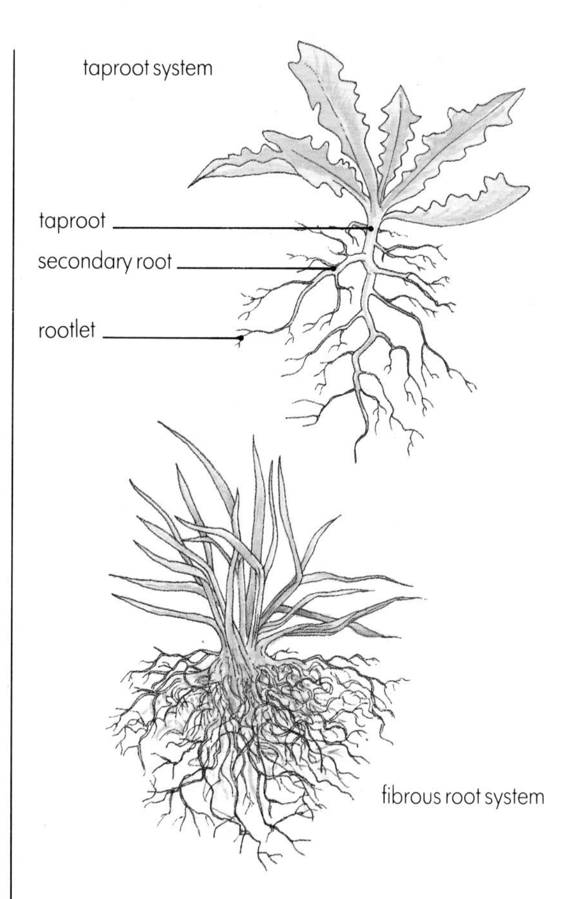

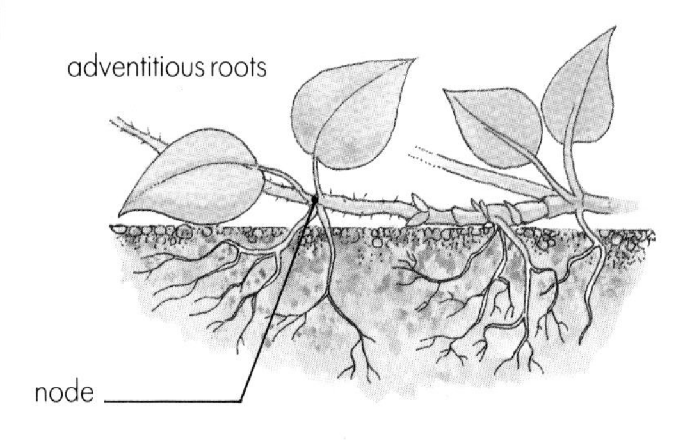

Above The three root types are fibrous roots, adventitious roots and taproots. All the other structures shown are specialized stems.

Below The scientific community represents flowering plants with diagrams depicting floral parts; (**left**) monocot, (**right**) dicot.

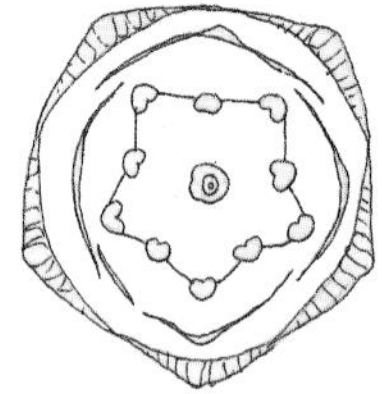

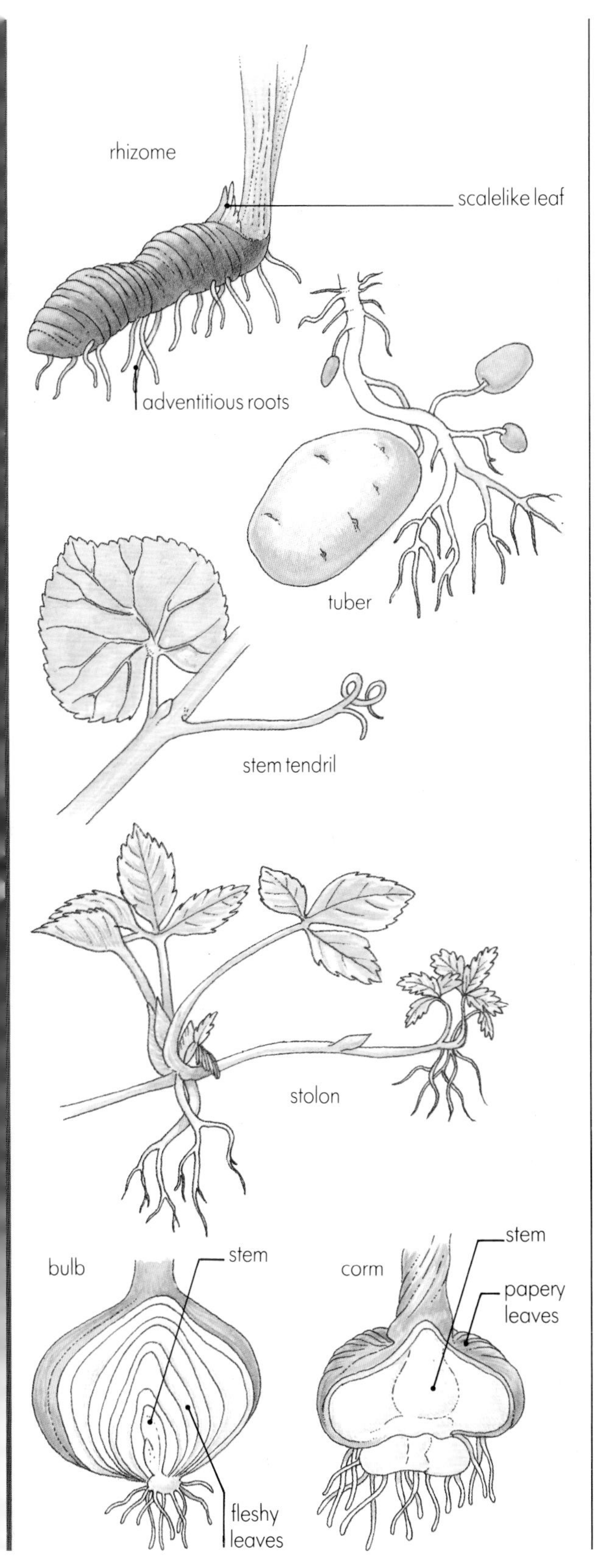

Above The conspicuous flower of yellow trillium, a southern and midwestern species, attracts insect pollinators; the leaf whorls on the stem allow maximum solar energy gain.

different species to fit into a wide array of ecological niches. They may form underground food storage organs, such as *rhizomes, bulbs, corms,* or *tubers.* Some plants form colonies by sending up *suckers,* or shoots, from rhizomes, or by sending out *stolons,* or runners, which put down adventitious roots upon reaching a suitable new location. Certain climbing plants, such as grapes and Boston ivy, produce *tendrils,* modified stems which extend from the parent plant to grasp other plants or objects for support.

Plants are woody or herbaceous depending upon the amount of *lignin,* a strengthening material, in the cell walls of their stem tissues. Woody, long-lived plants supporting great weight such as oak trees, produce a great deal more lignin than herbaceous plants, which support only a light weight for just one season.

Leaves

Although any green part of a plant can carry on photosynthesis, leaves are the primary sites of photosynthetic production. They often grow in a broad, flat shape, called the *blade,* which presents the maximum surface area to absorb the sun's rays. On many plants, the leaves twist on their stalks, or *petioles,* turning from east to west throughout the

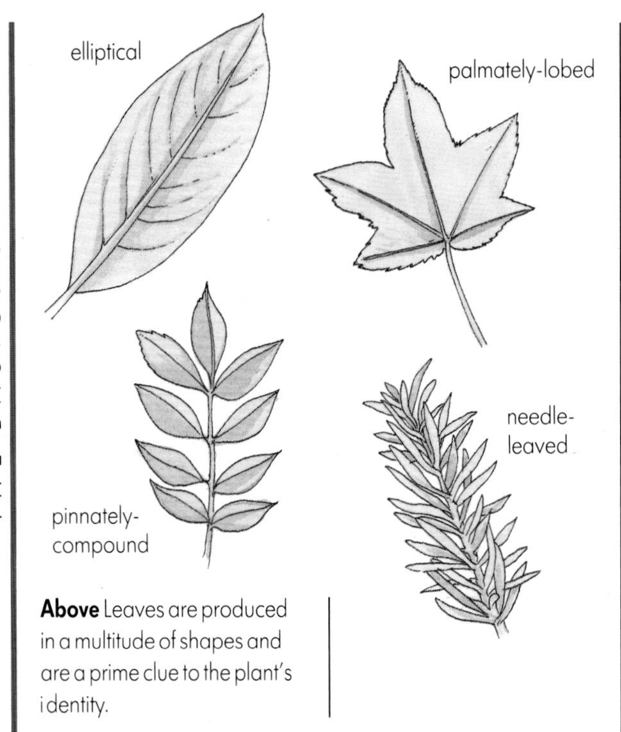

Above Leaves are produced in a multitude of shapes and are a prime clue to the plant's identity.

day to keep their leaves oriented toward the sun, returning to their east-facing position overnight to catch the first morning rays.

Leaves are usually arranged on the stem to take maximum advantage of the available sunlight. The largest leaves generally grow closest to the ground; leaves higher up the stem are progressively smaller, so they don't shade those below. Leaf arrangement may be *alternate,* with one leaf per *node* (point of attachment to the stem); *opposite,* with two leaves attached at the same node; or *whorled,* with three or more leaves per node. Leaves may be *simple* (undivided), or *compound,* with the blade divided into *leaflets. Pinnately compound* leaves have paired leaflets along a central stalk, or *rachis,* while the leaflets of *palmately compound* leaves radiate from a single point of attachment at the end of the petiole.

Leaves are important in the exchange of gases involved in photosynthesis and in *respiration,* an energy-releasing process. *Stomata,* tiny pores on the undersides of leaves, allow carbon dioxide to enter, and oxygen and water vapor, the by-products of photosynthesis and respiration respectively, to exit. Other waste products produced by the plant are deposited in the leaves and eliminated from the plant as its leaves are shed. Most water reaching the leaves evaporates, or *transpires,* into the atmosphere, resulting in a lower osmotic pressure in the leaves. The leaves then act as a siphon to draw water from the roots, relying on the cohesiveness of water molecules to pull a column of water to the tops of the tallest trees. In leaves, water and nutrients travel through vascular bundles called *veins,* parallel to one another in monocots, and in a branching network in dicots. Dicots may have *pinnately veined* leaves, in which secondary veins branch from a central vein called the *midrib,* or *palmately veined* leaves, where several main veins fan out from the base of the blade, like fingers on a hand.

Flowers

Flowers are reproductive structures composed of modified shoots and leaves. Their sole function is the production of fertile seeds which, in turn, produce the next generation. Pollination is usually accomplished by enlisting the aid of wind or insects, although some species rely entirely on nectar-feeding birds or bats, and a few are pollinated by water.

A flower begins at the tip of its stalk, or *peduncle.* This swells into the *receptacle,* to which all other parts of the flower are attached. A typical flower consists of the following parts: *sepals,* the outer whorl of leaf-like protective coverings of the bud; an inner whorl of *petals*; one or more *pistils,* the female organ consisting of a sticky, pollen-receiving *stigma* connected to the *ovary,* where seeds will ultimately develop, by a *style;* and several to many *stamens,* the male parts each consisting of a pollen-producing *anther* terminating in a slender *filament.* Collectively, the sepals are known as the *calyx,* the petals as the *corolla,* and the sepals and petals together become the *perianth.* There are a great many modifications to this arrangement in the plant kingdom.

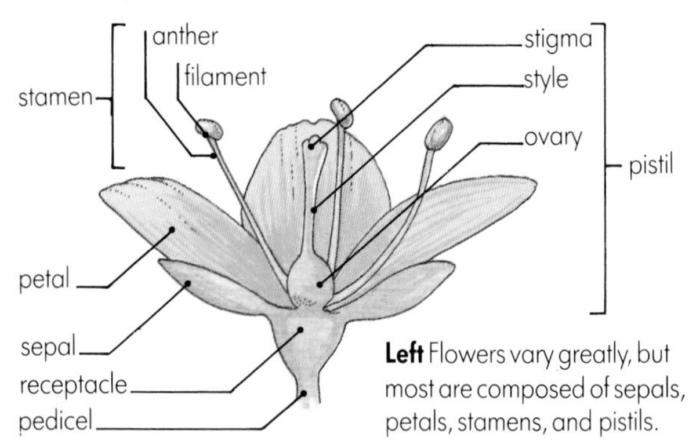

Left Flowers vary greatly, but most are composed of sepals, petals, stamens, and pistils.

What you can do

Dissect a flower

Larger members of the lily family work well for this. Observe an open bloom and try to identify its various parts. Carefully clip off a bloom near its base and, using a razor blade or a sharp knife, remove the sepals, petals, and stamens, leaving only the pistil still attached to the stem. Arrange the parts on a sheet of paper and examine them. (You can make a permanent record of your dissection by following the procedure outlined on page 33.) Once separated into its components, you can see that, despite the profusion of flower shapes, colors, and sizes, the underlying principle of a flower is really quite simple.

Left All flower parts are prominently displayed in this goldband lily, an Oriental species. Note the stamens with their pollen-laden anthers, the pistil, and the nearly identical petals and sepals.

Investigate xylem

You can visually demonstrate the function of xylem tissue with an ordinary white carnation. Tint some water in a glass (no more than six drops of food coloring per 8 ounces is necessary). Make a fresh cut near the bottom of the stem, half-way between the stem joints, or nodes, and place it in the water without immersing the bloom. In a few hours the bloom will have taken on the color of the water. You can have some fun by splitting the stem lengthwise into two or more segments and placing each in a different color. You can also perform this experiment using a plant with a light colored stem, such as celery, and actually see the tinted water progressing through the stem after about an hour. Try two pieces, one with leaves and one without, to see which one works faster. Can you explain your results?

Investigate roots and shoots

Roots orient toward gravity as surely as leaf stalks orient away from it. You can demonstrate both using easily germinated seeds such as tomatoes or radishes. Start the seeds between moist sheets of paper towel. When the roots are visible, mount several seeds between clear sheets of glass or plastic (lids from take-out salads work great), keeping a moist piece of towel between the seeds and one of the sheets. Use tape or a rubber band to hold the sheets firmly together. Stand the sheets on their sides until the roots elongate, then rotate the sheets 90° so that the roots and stem are horizontal. With further growth, you will notice that the roots have turned downward while the leaf stem has turned upward.

Plants accomplish this by means of *auxins*, chemicals which accumulate on the lower side of horizontal stems. Root cells are inhibited by auxins, and the faster growth of the upper cells forces the roots downward. Leaf stem cells are stimulated by auxins, so the more rapidly growing lower cells force the leaf stems upward.

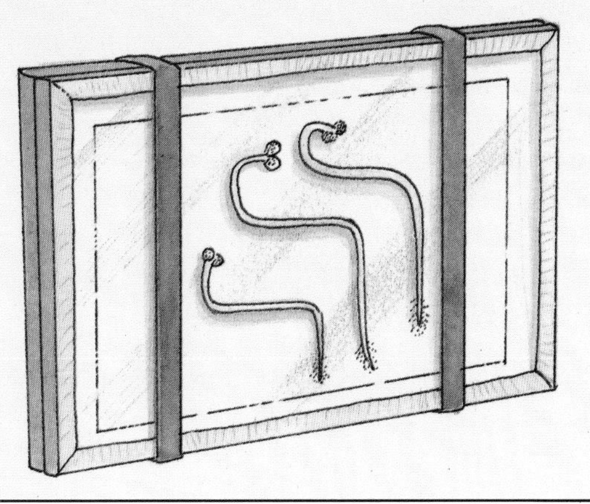

PHOTOSYNTHESIS: "EATING LIGHT"

We've mentioned that green plants are the only group of organisms capable of producing their own food, but how do they do it, and why can't we? Food is simply stored energy. More specifically, it is stored *chemical* energy. After we eat, our metabolism reduces the food to simpler forms and breaks certain chemical bonds, thereby releasing stored energy and nutrients and producing waste by-products. Energy released from food allows us to grow and also to carry on the activities necessary to maintain life.

Plants need not consume other organisms to obtain energy. Through their green pigment, *chlorophyll,* they are able to convert light energy to chemical energy and store it until needed. Through the action of chlorophyll and enzymes, a green plant can combine six molecules of carbon dioxide, six molecules of water, and light energy to produce one molecule of glucose, a simple sugar, and six molecules of oxygen, a by-product of photosynthesis.

Stored energy is of no use to an organism without some way to liberate it as needed. This is accomplished through *respiration,* a process taking place in all living cells 24 hours a day. *Aerobic respiration,* which occurs only in the presence of oxygen, is by far the most common form of respiration in both plants and animals. It is essentially the reverse of photosynthesis; one molecule of glucose and six molecules of oxygen are acted upon by enzymes to yield six molecules of carbon dioxide, six molecules of water, and energy. *Anaerobic respiration* occurs only under

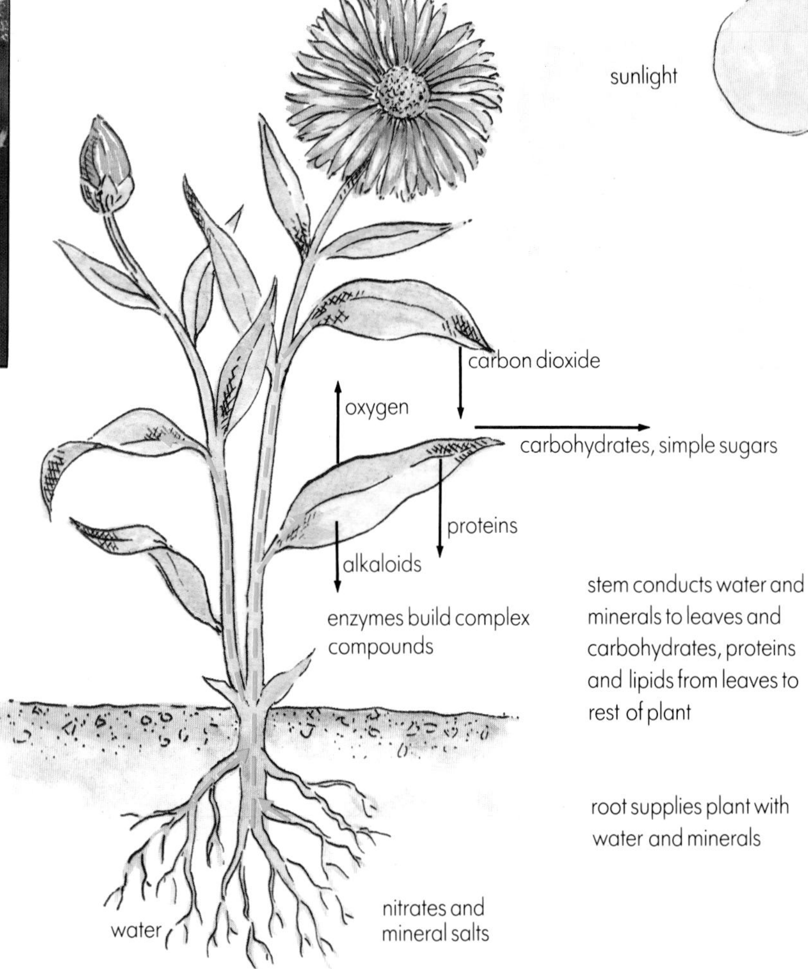

Above and **right** In most vascular plants, such as this scarlet oak, photosynthesis occurs primarily in the leaves. They are veritable food factories, receiving raw materials through stems and leaf pores and shipping finished products via the same routes. Energy liberated through respiration allows the leaves to convert the simple products of photosynthesis into more complex compounds useful in growth, reproduction, and defense.

special circumstances in the absence of oxygen; the most familiar example is fermentation which yields the ethyl alcohol in alcoholic beverages. Respiration is analogous to throwing a log on a bonfire; the stored energy in the log is transformed into heat and light, consuming oxygen and producing carbon dioxide and other gases in the process. In cellular respiration, the "bonfire" is miniscule and rigidly controlled, converting fuel to energy in tiny amounts as needed.

Some sugars produced by photosynthesis are used directly in respiration, but often the rate of photosynthesis is much higher than the rate of respiration, and more sugar is produced than can be used immediately. Sugar not needed for respiration is transformed into proteins, lipids (fats) or other carbohydrates, such as cellulose, sucrose, and starch. At peak periods of photosynthesis, excess water-soluble sugar is transformed into insoluble starch, so as not to upset the osmotic balance of the cell, and some is stored in the *chloroplasts,* the chlorophyll-containing organelles (microscopic structures) of plant cells. It is later broken down into smaller soluble sugars, and transported out of the leaf for use or storage elsewhere in the plant.

What you can do

Demonstrate photosynthesis

We cannot directly observe photosynthesis, but we can observe its by-product. Collect a few submerged aquatic plants, such as pondweed, from a pond, or purchase some from a tropical fish store. Place a water-filled aquarium or a large glass container, such as a pickle jar, in a sunny location. Submerge the plants under an upturned glass, making sure no air remains in the glass. Over several days, you should observe an increasing accumulation of oxygen at the top of the glass; you may even observe bubbles rising from the plant. By marking the oxygen level at the end of each day and making careful observations, you can co-relate oxygen production with the number of hours of sunlight each day. You can manipulate the amount of light reaching the plant by using an aluminum foil reflector, a special plant light, or by draping a dark cloth over the aquarium. Corresponding increases or decreases in oxygen production will be noted with such modifications.

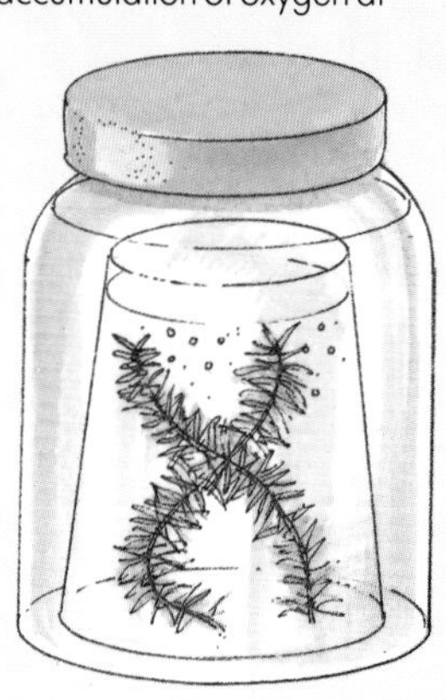

Demonstrate osmosis

Photosynthesis is possible in terrestrial vascular plants because water is absorbed by the roots from the soil through the process of *osmosis*. Osmosis is the movement of water through a cell membrane from a greater concentration of water to a lesser

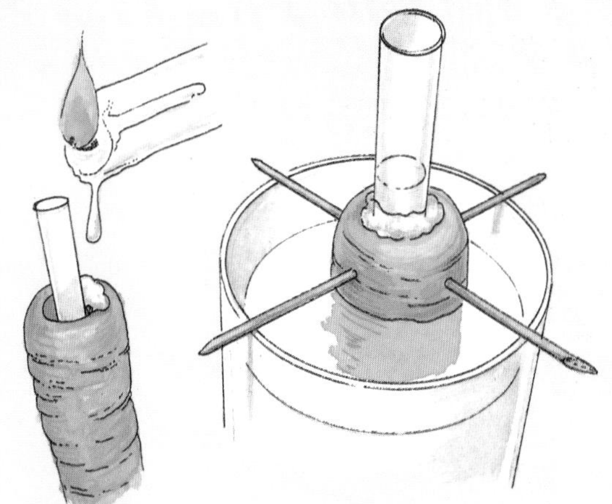

concentration. You can make a device to demonstrate osmosis in roots using a carrot and a plastic drinking straw. Ordinarily, a carrot growing in the ground absorbs water by osmosis and transports it to the stems and leaves. If you remove the stems, hollow out the top of the carrot, and insert a clear straw, you can observe water rising in the straw by osmotic pressure. Simply seal the straw in the carrot with candle wax, drip a small amount of sugar water down the straw and mark its level, support the carrot in a glass of water using toothpicks, and observe. After a short period, you will notice that the fluid level in the straw has risen as the water flowed toward the sugar water (the lesser concentration of water, since sugar occupied part of the volume). Tinting the water with food coloring or ink and then slicing the carrot lengthwise at the end of the experiment will reveal the veins through which water travels, as well as the untinted cells which store food.

LIFE CYCLES

The lives of plants, like those of all other organisms, are finite, having both a beginning and an end. The interval in between ranges from days to centuries, depending upon the species. Plants hold the record for longevity on earth; the General Sherman, a 272-foot giant sequoia tree in Sequoia National Park, California, is more than 3,000 years old, an age surpassed only by a few bristlecone pines.

The more advanced groups of plants can be divided into three general categories based on their life cycles. *Annuals* complete their entire life cycle in one growing season. *Biennials* require two seasons to complete their life cycle; in the first season they are vegetative, and only flower and reproduce in their second year of life. *Perennials* continue to live and flower for some years after reaching maturity.

The beginning of a flowering plant's life cycle is preceded by the pollination of its parent. *Pollination* is the transfer of *pollen*, containing the plant's sperm, from the anther to the stigma. When this occurs between the flowers of different plants it is called *cross-pollination*, a process favored among flowering plants because it promotes a random mixing of genes within the gene pool of the

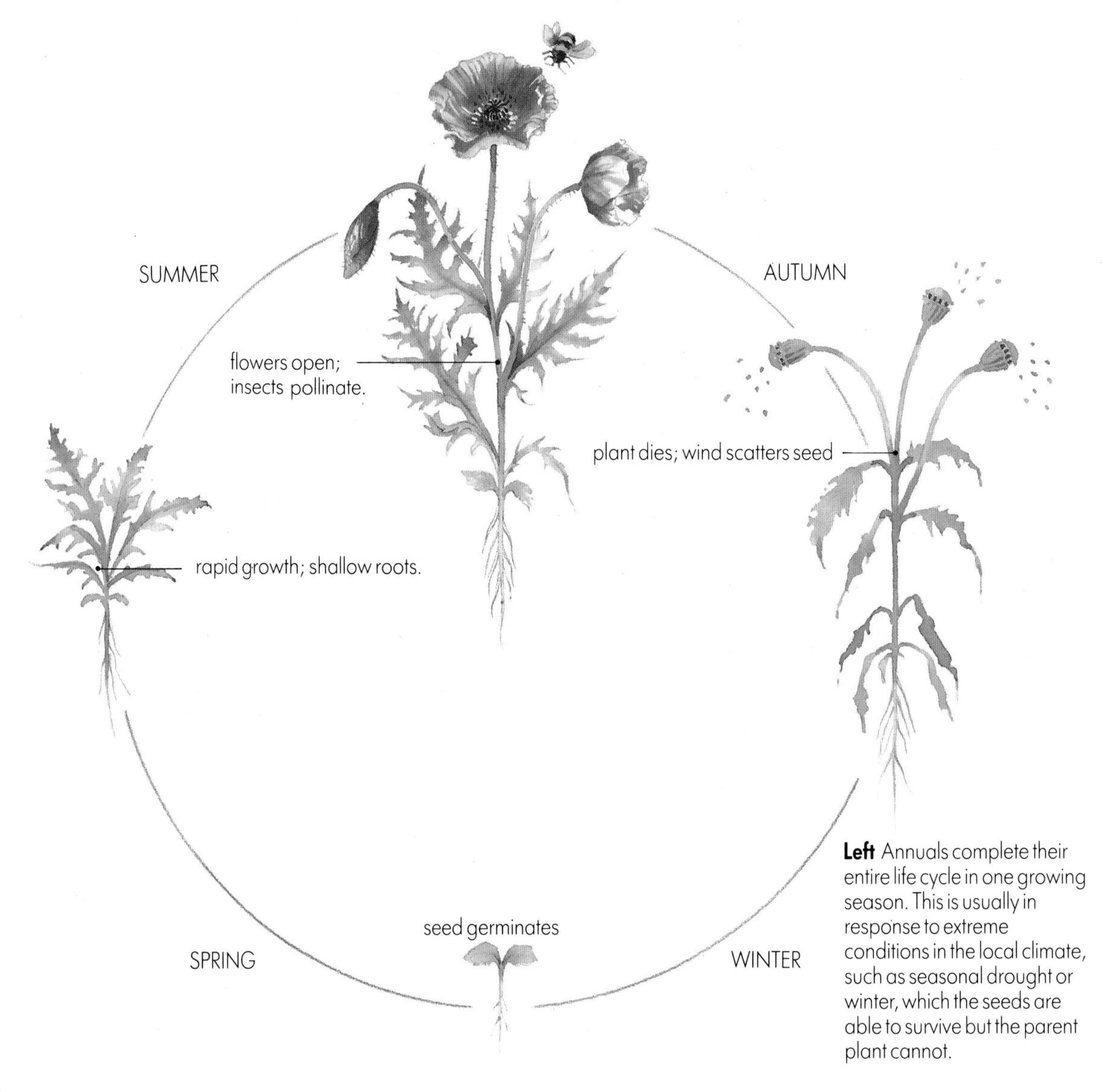

Left Annuals complete their entire life cycle in one growing season. This is usually in response to extreme conditions in the local climate, such as seasonal drought or winter, which the seeds are able to survive but the parent plant cannot.

Wind-pollinated flowers produce copious quantities of light, airborne pollen and vast numbers of female flowers, relying upon the whims of air currents to unite the two. On the other hand, insect-pollinated flowers produce heavy, sticky pollen grains designed to adhere to a visitor's body. Insect-pollinated goldenrod (**right**) is often erroneously blamed as the cause of hayfever, but the real culprit is usually the less conspicuous, wind-pollinated common ragweed (**above**).

species. The resulting diversity of traits better enables that species to cope with changing environmental conditions, which is crucial to the process of evolution.

Plants employ many methods to ensure that cross-pollination occurs. Some produce plants of separate sexes, so that cross-pollination is inevitable. Others develop stamens first, and delay maturation of the pistil until the stamens have ripened and shed all of their pollen. However, many species produce both stamens and pistils simultaneously. How do they promote cross-pollination? Most plants in this category have flowers designed so that the pollinator, usually an insect or a nectar-feeding bird, comes into contact with the stigma before touching the anthers, so that pollen transferred is more likely to come from another plant.

Flowers are usually wind pollinated or pollinated by animals. Wind-pollinated flowers are inconspicuous, since they don't have to attract pollinators. Because wind is unpredictable, these

Above Plants have evolved a myriad ingenious ways of dispersing themselves to suitable new habitats. Many involve seed adaptations, such as the parachute-like appendages of dandelion seeds, allowing them to ride great distances on air currents.

plants produce pollen in astronomical quantities, relying on sheer numbers to get the job done. Their buoyant pollen grains are often carried great distances on the wind, and are responsible for the hayfever allergies suffered by millions of people. Animal pollination, on the other hand, requires the flowers to be conspicuous to potential pollinators, and there would seem to be nearly as many combinations of form, color, and fragrance as there are pollinating species. The greatest number of pollinators are insects, so many floral features, such as distinctive color combinations, bold markings called "honey guides," which lead insects toward the flower's nectar and/or pollen, and specific fragrances all lure potential pollinators. Insects perceive ultraviolet as a color, but most cannot see red, so, many insect-pollinated flowers incorporate ultraviolet features, while all-red flowers rely on hummingbirds and some species of butterflies to accommodate them.

Once pollination occurs, the next step is fertilization. Pollen deposited on the sticky stigma generates a fine *pollen tube* that conveys the sperm through the style to the ovary, where the ovules, or eggs, have developed. After fertilization, the rest of the flower parts wither and are shed as the ovary swells with seed development.

When seeds mature, they face the formidable task of dispersing to suitable sites for germination – no small feat for organisms without apparent means of locomotion. Dispersal is particularly important for seeds of long-lived plants which would face overwhelming competition from their mature parent if they were to simply fall and germinate at its base. Plants have, however, evolved a multitude of ingenious methods to ensure that at least some of their seeds reach locations where the seedlings can compete. Plants such as dandelions, milkweeds, and thistles rely on airborne dispersal, producing downy seed "parachutes" that ride air currents. Some plants hitchhike on animals to reach their destinations. The seeds of common

burdock, for example, are equipped with hooks which cling tenaciously to fur or clothing until knocked or pulled off, usually some distance from the parent plant. Another means of hitchhiking is employed by seeds that develop inside a palatable fruit or berry, which is eaten by a bird or mammal, then passes through the animal's digestive tract, emerging some distance away. Seeds of many wetland plants float on water, while those of others, such as spotted jewelweed, are shot out of spring-loaded pods.

The interval between seedling and maturity varies greatly among species. Annuals grow rapidly, putting down only shallow roots and concentrating their resources on producing vegetative growth and flowers. In contrast, herbaceous perennials, trees, and shrubs, invest heavily in well developed root systems, often postponing flowering for several seasons. Some are opportunists. Many trees, for example, spend years as seedlings in a mature forest, patiently waiting for one of the old monarchs to fall, then race to fill the hole in the forest canopy with their own branches before their competitors can do so.

Above Another seed dispersal adaptation is the hitch-hiking technique employed by many plants, including common burdock. Hooked extensions snag the fur, clothing, or feathers of passers-by, and the seeds remain attached until they are worked free, usually at some distance from the parent plant.

What you can do

Discover hidden seeds

During the next spring thaw, put on your boots and take a walk someplace muddy. When you get home, scrape the mud from your boots and mix it with water to a semi-liquid consistency, and leave overnight. The next day, add the mud-and-water mixture to sterile potting compost, cover with a sheet of glass or clear plastic, and place somewhere warm and light, keeping the compost moist but not too wet. In two or three weeks, seedlings should emerge. As they grow, you can transplant them and raise them to maturity to identify them and observe their life cycles.

SEASONAL CHANGES

Temperate-region plants must cope with extremes in climate through the year. To survive, they have evolved a great array of adaptations. Annuals simply limit themselves to one growing season, during which they sprout when conditions are prime, grow, flower and set seed. Seeds are quite hardy and can withstand the rigors of winter while awaiting conditions favorable for germination and growth. In fact, many species' seeds must undergo an extended period of freezing to germinate. This prevents their sprouting during a mild "Indian summer," when conditions often mimic spring.

Biennials must weather one winter, and usually spend it as a low-growing plant, sheltered from the elements under the debris of expired vegetation and a blanket of snow. Not until their second season do they rise to a mature height and flower.

Perennials are truly innovative when it comes to coping with the seasons. Herbaceous perennials often store food in modified underground stems before discarding their above-ground growth and entering a dormant period. Evergreen plants produce a resinous sap, a sort of botanical antifreeze, so their leaves survive winter without freezing, and are ready to begin photosynthesis as soon as the ground thaws and water becomes available.

Deciduous plants

Deciduous plants drop their leaves seasonally. Evergreen plants also shed their leaves, but much less conspicuously, dropping them sporadically throughout the year. The process by which leaves

Below Herbaceous perennials, such as this wind-blown goldenrod, cope with winter by storing food in the roots and dying back above ground.

Above and **right** The shedding of leaves by deciduous trees, shrubs, and vines enables them to avoid the stresses of winter. As the green chlorophyll begins to diminish, yellow pigments are exposed. Sugar trapped in the leaf is converted into red, blue, and purple pigments which, when combined with the yellows and greens, result in the brilliant display of autumn colors familiar in the Northeast.

are shed is called *abscission.* In deciduous plants, hormonal responses to environmental changes such as decreasing day length, light intensity, temperature, and precipitation trigger the formation of specialized cells in the *abscission zone* at the base of the petiole. These cells are easily broken down by plant enzymes, leaving the leaf attached to the stem only by a few strands of xylem. Wind and rain easily sever this connection, and the leaf falls, leaving on its twig a *leaf scar* encircling several *bundle scars* where the vascular bundles of xylem and phloem once entered the leaf. A corky layer below the abscission zone prevents water loss from the leaf scar.

Preceding deciduous leaf fall is the color change that has become the hallmark of autumn. Chloroplasts, the cell structures in which photosynthesis takes place, contain groups of pigments including chlorophylls (green), carotenes (yellow or orange yellow), and xanthophylls (pale yellow). Chlorophylls are present in the greatest quantities in mature leaves, and their color masks other pigments. As cells in the abscission zone deteriorate and die, they begin to close off the vascular bundles that transport materials to and from the leaf. Without this influx of raw materials, chlorophyll is not replaced, and its diminishing concentration allows other pigments to show through.

The clogged vascular bundles also prevent the leaf's sugars and waste products from leaving. Trapped sugars are converted into anthocyanins, the deep-red pigments responsible for many autumn hues. The exact autumn color of a leaf often depends upon the acidity or alkalinity of the *protoplasm,* or sap, in its cells. Anthocyanins will turn red if the protoplasm is acidic, blue if alkaline, and purple if neutral. Such conditions vary from cell to cell, hence the kaleidoscope of colors.

A combination of bright, sunny days and cool nights, coupled with ample moisture, produce the most colorful autumn scenes. Bright sunshine stimulates the leaves to continue producing sugars at a rapid rate, and cool nights, around 40°F, cause the abscission zone to pinch off the vascular bundles quicker, trapping more sugar in the leaves. Dry weather diminishes the intensity of fall colors, because parched leaves produce less sugar, and therefore less anthocyanin.

Once deciduous leaves are shed and the sap has retreated to its root system, a woody plant becomes dormant. On the twigs, together with the leaf scars and bundle scars, are buds, the birthplace of the next season's growth. Buds are formed toward the end of the previous growing season. *Lateral buds* develop in the leaf axils, the angle between the stem and leaf petiole. On a winter twig they are located immediately above the leaf scars. The *terminal bud* forms at the end of the twig, and marks the end of last season's growth. Buds are covered by at least one, but usually two or more, protective bud scales. Larger buds enclose flowers or leaves and flowers, while smaller buds contain only leaves. In woody plants, all buds produce wood growth in addition to flowers and/or leaves. Also located within each terminal bud is the *apical meristem*, a region of actively dividing cells. Though dormant during winter, the apical meristem reactivates in spring and will produce the following year's growth. Below the terminal bud are several

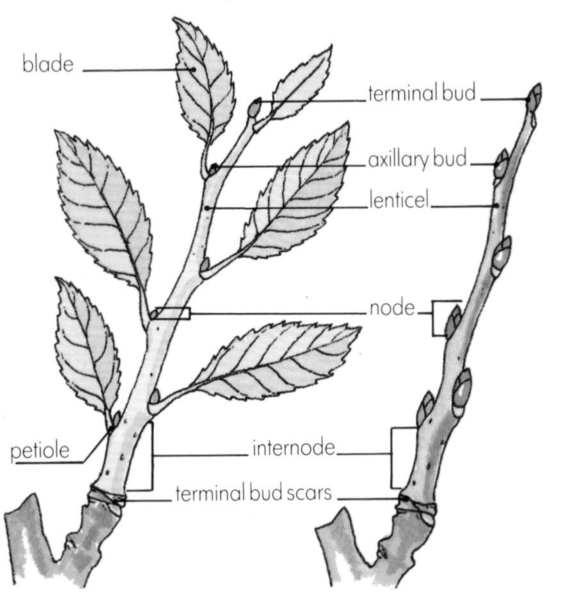

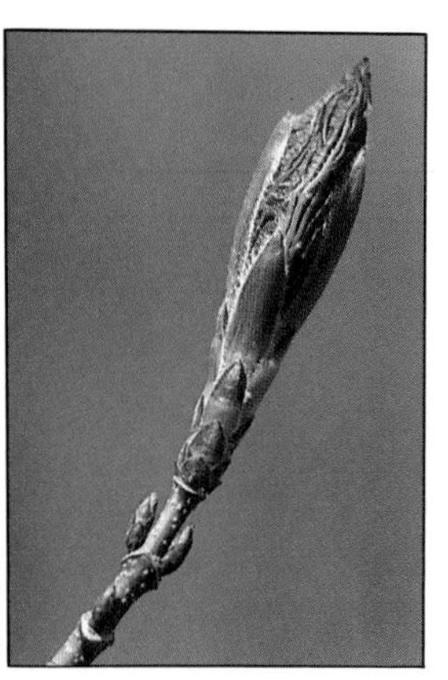

Left and **above** Buds enclose the tissues that produce the following year's growth. Encased within protective bud scales is the apical meristem, which produces new woody tissue, as well as a compact flower and/or leaf. In the spring, these tissues swell with new sap and force their way out of a bud, such as this sugar maple.

What you can do

Fool mother nature

Dormant tree twigs can be brought indoors to a warm room and their buds "forced" open by placing the twig in water. Wait until after December, since many Temperate-climate trees need a period of cold for their buds to open, ensuring that they do not mistakenly open during warm autumns. Many insect-pollinated trees such as serviceberry, apple, and cherry produce impressive flowers, though the less conspicuous blossoms of wind-pollinated species, including oak, birch, maple, and willow, are also interesting, and the intricacy of unfolding leaves is a marvel to behold. You may wish to dissect an unopened bud with a razor blade, and try to recognize any of the structures within before they emerge.

scars encircling the stem, marking the point of attachment of last season's bud scales. Each interval between bud scale scars represents one year's growth. By counting these scars until you reach the parent branch, where the twig originated from a lateral bud, you can determine its age.

Another seasonal change illustrating periods of growth and dormancy are the annual rings on cut timber. Cambium, a layer of cells similar to those of the apical meristem, allows for the expansion of the trunk's girth to support the increasing weight resulting from stem elongation. New xylem tissue is laid down inside the cambium layer, and new phloem on the outside. When the cambium is first activated in spring, it forms relatively large, tubular *vessel elements* of xylem, the light-colored "spring wood." Later in the season, smaller vessel elements and more *tracheids,* a porous, non-tubular type of xylem tissue, are produced, resulting in darker-colored "summer wood." Both vessel elements and tracheids become impregnated with lignin and die soon after their formation, but they remain intact and functional as transport pipelines. The concentric circles of light and dark wood that result are called *annual rings,* each composed of one layer of spring wood and one layer of summer wood. Counting annual rings will accurately yield the tree's age when cut, and can often reveal much about the life of the tree. Events such as drought, fire, infestations, injury, and harvesting of surrounding timber all result in altered growth patterns.

Above Trees grow in girth as well as in height. A specialized layer of cells, the cambium, produces a new layer of vascular tissue each growing season, resulting in the annual rings visible on the end of cut timber. By counting these rings, you can determine the exact age of the tree.

TAXONOMY: SIMPLICITY FROM CHAOS

Taxonomy is the science of identifying, naming, and classifying organisms. It is the discipline which puts order into an immensely diverse world and allows scientists to discuss any organism in the certainty that they are talking about the same one.

Common names are generally used in everyday conversation, but may not positively identify a particular species. Many plants and animals have more than one common name, and may be known by different names in different areas, and the same common name may be applied to different species. Clearly, the potential for confusion is great, with well over 350,000 plant species known and many more remaining to be discovered.

Contemporary scientists use a classification hierarchy to organize life forms into a series of categories arranged in order from general to specific relationships. They are, in order of increasing specificity: kingdom, division (or phylum, in the animal kingdom), class, order, family, genus, and species. Each of these is a collective unit composed of one or more groups from the next lower category. For example, a genus is a closely related group of species, and a family is a closely related group of genera.

While classification has always been a fairly simple affair, naming has not. By the beginning of the eighteenth century, the use of Latin in schools and universities was widespread, and it was customary to use descriptive Latin phrases to name plants and animals. All organisms were grouped into genera, and the descriptive phrase began with the name of the appropriate genus. All known mints, for example, belonged to the genus *Mentha*. The complete name for peppermint was "*Mentha floribus capitatus, foliis lanceolatis serratis subpetiolatis*, or "Mentha with flowers in a head, leaves lance-shaped, saw-toothed, with very short petioles." The closely related spearmint was named *Mentha floribus spicatis, foliis oblongis serratis*, which meant "Mentha with flowers in a spike, leaves oblong and saw-toothed." Though quite specific, this system was much too cumbersome to use efficiently.

In 1753, the Swedish naturalist Carolus Linnaeus introduced a two-word system of naming organisms. It quickly replaced the older, clumsier method, and came to be known as the Binomial System of Nomenclature. It identifies individual species by linking the generic name with another word, frequently an adjective. All scientific names are Latin, although some have descriptive Greek roots. The first name is always capitalized, but never the second, and both are either underlined or italicized. When more than one member of the same genus is being discussed, once the generic name is given in full, it may be abbreviated when referred to again, as in *A. rubrum* for *Acer rubrum*.

Caltha palustris (**above**) and *Calta leptosepala* are both known as marsh marigold. Common plantain, *Plantago major* (**right**) is known by at least 45 other names in English alone.

TREES AND SHRUBS

Technically, a shrub is a woody perennial branching at or near ground level, while a tree is a woody perennial with one main stem branching some distance above the ground. However, many contemporary field guides classify some small trees as shrubs, and some trees are naturally multi-stemmed, so the difference is somewhat arbitrary, but both have permanent woody frameworks that do not die back in winter.

The vast majority of trees and shrubs can be placed into one of two basic divisions. Division Coniferophyta, or conifers, includes needle-leaved trees that do not flower but produce their seeds in cones: pines, spruces, firs, hemlocks, cedars and redwoods, for example. The largest group of plants, Division Anthophyta, the flowering plants, includes most broad-leaved trees and shrubs. In addition, all

Left The quaking aspen is easily identified by its round, fine-toothed leaves which tremble in the slightest breeze.

Top and **above** The needle-like leaves of conifers, such as the eastern white pine and pitch pine, enable them to withstand extremes better than most deciduous trees.

Above Shrubs, such as this flame azalea, are flowering woody plants, deciduous or evergreen, with several stems branching at or near the ground. Certain small trees with one main trunk may also be categorized as shrubs.

trees and shrubs in this division belong to Class Dicotyledonae, the dicots.

Like most plants, trees do not grow randomly, but need precise conditions to flourish, and when they find these conditions, grow profusely. Forest habitats are identified by their one or two dominant tree species, such as an oak-hickory forest or a boreal forest, which is composed primarily of black spruce, white spruce, and balsam fir. Naturalists know what type of conditions prevail in an oak-hickory forest, so they can also predict with some certainty the other plants, animals, climate, soil factors, and possibly even the landforms likely to be found there.

Trees and shrubs differ from herbaceous plants not only in their production of woody stem tissue, but also in their ability to produce bark, the protective outer "skin" that enables them to survive from decades to centuries. Inner bark refers to the phloem laid down by the cambium layer, while outer bark, the tough protective covering of cork tissue, is produced by a layer of cells immediately outside the phloem called the cork cambium. Bark reduces water loss and protects the stem against mechanical injury, bacterial or fungal infection, or infestation. In areas where fire is relatively common, some species have adapted to the threat by producing bark that is fire-resistant, and in some cases, nearly fire-proof.

What you can do

Take bark rubbings

Making bark rubbings enables you to "collect" bark samples to study, and to compare the barks of different species, even those that do not grow near one another. You can add a bark rubbing to your record of a tree species, which may already include sketches, photographs, specimens of leaves and fruit, or notes on interesting or unique features. If possible, try to get a rubbing of one of the lower branches, which often have different patterns and textures to that of the trunk.

You need sticky tape, a large wax crayon, and strong but thin paper. Tape a piece of paper in position. Using the flat side of the crayon, rub slowly, making all strokes in the same direction. For future reference, label your rubbing with the tree's common and scientific names, your name, the date, and the tree's location.

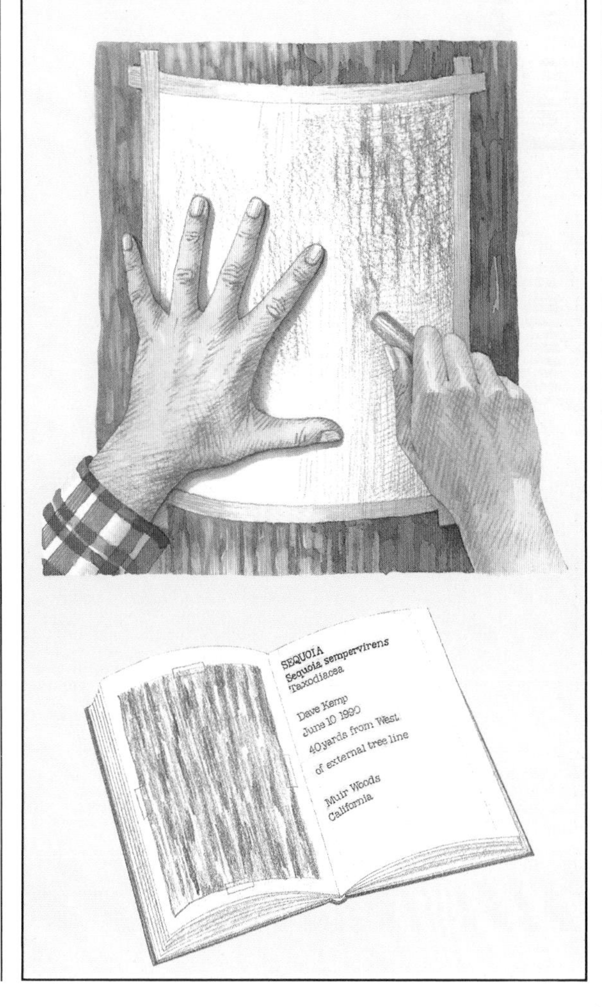

WILDFLOWERS AND GRASSES

Wildflowers are another arbitrary category of plants. All are, of course, members of the Class Anthophyta, the flowering plants, and include both monocots and dicots. By general definition, they grow without cultivation. Some have woody stems, but most are herbaceous. To complicate matters, certain shrubs are sometimes classified as wildflowers in field guides because of their blooms.

Wildflowers, like trees and shrubs, are a diverse group, with seemingly infinite combinations of flower and leaf shape, structure, color, arrangement, and other features. Flowers may be *radially symmetrical* (wheel-shaped, capable of being divided in half along more than one lengthwise plane), or *bilaterally symmetrical* (capable of being divided in half by only one lengthwise plane). The buttercup is radially symmetrical; the lady's slipper is bilaterally symmetrical. Some produce only one bloom, while others produce hundreds in one compact cluster. Their leaves may be *simple* (undivided), or *compound* (divided into *leaflets*). A large number of leaf shapes occur, such as *ovate* (egg-shaped), *lanceolate* (lance-shaped), *linear* (long and narrow), and others. Leaf *margins* (edges) may be *entire* (smooth), *lobed, undulate* (wavy), *serrate* (toothed), or a combination of these.

How does one put order into such an eclectic group of plants? A good field guide helps in identification but the real key is to become familiar with plant families. Families are handy divisions for learning about wildflowers because they are large enough to include plants with general similarities, yet small enough not to be vague and cumbersome.

Below Black-eyed Susan and other members of the sunflower family produce composite flowers consisting of many disk and ray florets.

Radially-symmetrical or regular, flowers like purple trillium (**above**) are wheel-shaped and can be divided into two identical parts by numerous planes. Bilaterally-symmetrical, or irregular, flowers such as pink lady's slipper (**left**), can be divided into two mirror-image parts by only one plane.

Left One of the taller members of the grass family, giant reed is common in wet areas, especially in or near brackish water. Like all grasses, it flowers and produces seeds, but its primary method of reproduction is via spreading rootstocks.

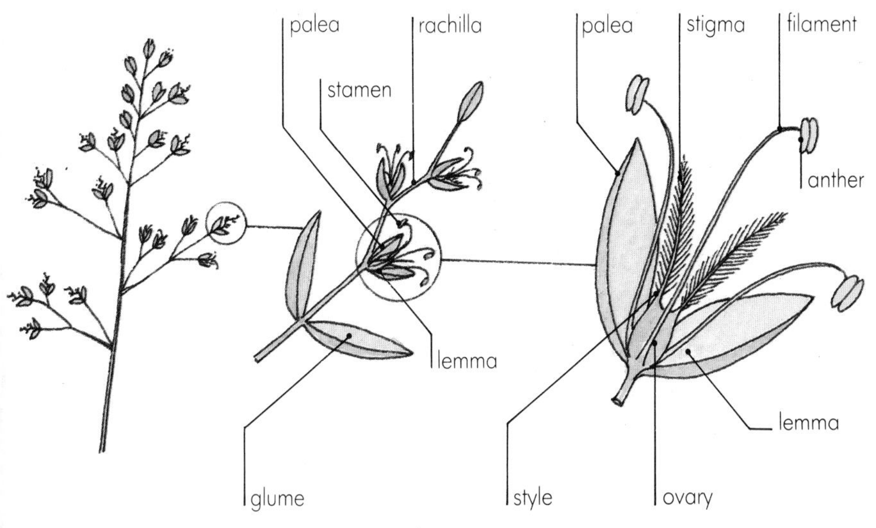

Left Many people do not recognize grasses as flowering plants because their flowers are often small and inconspicuous. Being wind-pollinated, they do not need large or showy blooms to attract pollinators.

More than one hundred families of wildflowers exist in temperate North America, but most species fit into about two dozen or so families. Learning the family traits listed in your field guide allows you to examine a flower and narrow the possibilities from many thousands to a small handful.

One family of flowering plants stands above the rest in terms of its significance to humanity: the grass family *Poaceae*. Although there are only about one-quarter the number of species in the grass family as there are in the sunflower family, *Compositae*, grasses are by far the most widely distributed family and the largest in terms of individual plants. They include all the cereal grains, such as corn, wheat, oats, rice, barley, and rye. In fact, nine of the ten most economically important plants in the world are grasses. More than one billion tons of cereals are harvested annually to supply more than half of the food needs of the world's human population.

Grasses are monocots, a class that represents approximately one-quarter of all flowering plants and also includes rushes, sedges, orchids, lilies and irises. They exhibit typical monocot traits, including parallel-veined leaves and flower parts in multiples of three. Grass flowers are highly specialized, unlike the rushes, another grass-like family with a more conventional flower structure. The calyx and corolla of grasses are reduced to tiny, inconspicuous scales, and the feathery stigmas effectively sift air-borne pollen from the wind. Grass leaves sheath the stem at their bases. Grass stems are cylindrical and hollow, major features distinguishing them from sedges, which are similar in appearance but have solid stems, triangular in cross-section.

MOSSES, LIVERWORTS, FERNS, AND CLUB MOSSES

Mosses and liverworts belong to the Division Bryophyta. Bryophytes are the most primitive members of the plant kingdom. With no vascular system to supply water to their extremities, they must absorb water directly through their surface from their environment, and are hence low-growing plants of moist places never more than a couple of inches high. Most are soft and pliable, since they lack xylem and phloem tissue. Bryophytes are like amphibians, in that water is essential for fertilization to take place. However, only a thin film of water is needed for the plant's sperm to swim and find the egg.

Liverworts

"Wort" means plant or herb. Liverworts are so named because some members of this group, the Class Hepaticae, grow as flattened lobes called *thalli*, somewhat resembling a human liver. Such lobes are common in only about one-fifth of the liverwort species, however. The rest are more "leafy" and superficially resemble the true mosses, except that liverwort "leaves" are scaly, and grow in two rows on a prostrate "stem", whereas the "leaves" of mosses grow spirally around an upright "stem". The lower surfaces of thalloid liverworts have numerous single-celled *rhizoids* that resemble tiny roots and function as anchors. Liverworts are particularly abundant in moist places where the light level is too low for competing flowering plants. Identification of individual species can be quite difficult.

Mosses

Mosses are members of the Class Musci. Like liverworts, mosses only grow in moist places, or in areas that are damp part of the year. Moss "leaves" are attached to an axis, the "stem," which is anchored at the base by rhizoids. Mosses, like liverworts, must absorb water directly through their surfaces. Since the upright stems possess no xylem, they depend on water flowing up the outside of the plant by means of capillarity, the ability of a thin film of water to flow up a surface due to the tendency of water molecules to cling together.

Mosses are divided into three subclasses: peat mosses, true mosses, and rock mosses. By far the best known of these is a remarkable group of peat mosses belonging to the genus *Sphagnum.* Sphagnum mosses grow in very moist areas such as around the perimeter of lakes and on floating mats of vegetation in bogs. Like all mosses, they act as environmental sponges, absorbing water quickly and releasing it slowly, thus reducing flooding and erosion. Sphagnum mosses are particularly adept at this; a pound of dry peat moss will absorb 25 pounds of water. This property makes it an excellent soil conditioner in gardening, and it also adds *humus*, or organic

Left Common throughout northern North America, sphagnum mosses are found in bogs, along the shorelines of lakes and ponds, and on damp forest floors.

Right Liverworts are another non-vascular plant. With no xylem tissue to transport water throughout the plant, liverworts are confined to damp locations where they can maintain a thin film of water over the plant's surface.

matter, to the soil. Also known for its antiseptic properties due to its natural acidity, sphagnum moss has been used as a poultice on wounds at least as far back as the Crimean War of 1854-1856.

Club mosses

Club mosses are not really mosses. They belong to the Division Lycophyta, primitive non-flowering vascular plants, and as such differ from true mosses, which have no xylem or phloem. The two major genera are *Selaginella* and *Lycopodium*. Species of *Lycopodium* are also known as ground pines because they grow in forests and look superficially like tiny conifers. Their upright stems, less than one foot tall in temperate regions, grow from branching rhizomes which put down adventitious roots as they go. Club mosses reproduce asexually from spores produced on *strobili*, cone-like projections on the tops of plants, but they also have a sexual generation that is usually hidden below ground level.

Ferns

Ferns have an undeserved reputation of being difficult to identify, and so are often ignored by novice naturalists in favor of wildflowers and trees. Though there are more than 11,000 fern species, the vast majority grow in the tropics, and it is simpler to learn the few dozen fern species found in temperate North America than the 10,000+ species of wildflowers in the same region.

Ferns are fairly advanced among non-flowering plants, having true roots, stems, and leaves. Their stems possess xylem and phloem vascular tissue, enabling them to grow to substantial heights. Many ferns spread by underground rhizomes, resulting in dense colonies of a single species. Fern leaves, called *fronds*, are often divided into leaflets, or *pinnae*, which themselves may also be divided, resulting in a lacy appearance. Henry David Thoreau, the nineteenth century author and naturalist, once stated that "God created ferns to show what he could do with leaves." Fern fronds first appear as tightly coiled "fiddleheads," which then unfurl and expand into the blades.

Like mosses, liverworts, and club mosses, ferns alternate between sexual and asexual reproduction and depend upon a thin film of water for the sexual portion of their life cycle. Mature fronds produce spores asexually in clusters called *sori*, which in many species are protected by individual flaps of tissue called *indusia*. Their pattern of development on the frond is often a key to identification. A wind-borne spore that lands on a favorable site produces root-like hairs anchoring it to the soil, and proceeds to grow into a small, flattened, heart-shaped plant known as a *prothallus*, the sexual portion of its life cycle. Regardless of the number of eggs fertilized, only one spore-bearing plant will grow from each prothallus, thus completing its life cycle.

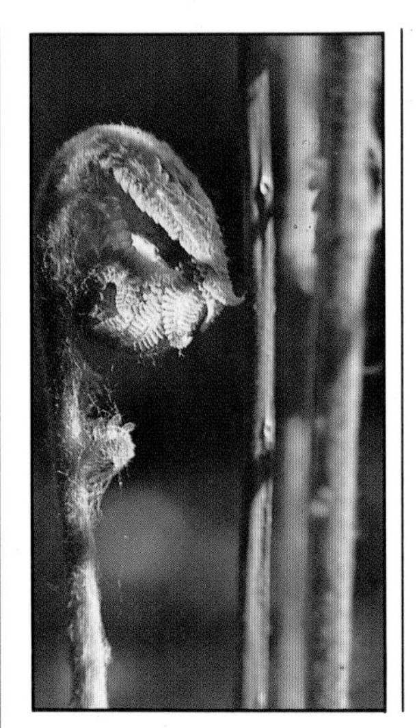

Left Ferns begin their asexual (spore-bearing) generation as a tightly coiled structure commonly known as a fiddlehead. A familiar spring sight in many areas, fiddleheads arise from underground rhizomes, gradually unfurling into a frond. Certain ferns are evergreen, while others produce new fronds every year.

Left Club mosses are common inhabitants of North American forest floors, and are usually found poking through the leaf litter.

Above Many ferns produce sori, or spore clusters, in patterns that vary according to species, on the undersides of fronds. Each sorus may be covered with a protective flap of tissue, called the indusium.

What you can do

Preserve mosses and liverworts

Mosses and liverworts contain lots of water, and must be dried thoroughly to preserve them. Place the specimen in a cheesecloth sack and hang it in a warm, dry area. At room temperature, the plant should be dry in 7-10 days. Store in a box or display case to prevent damage to delicate structures such as spore capsules. Place a small muslin bag of silica gel in the same box to help preserve your specimen. Dry the silica gel in an oven periodically to maintain its effectiveness. Label your specimen with its name (if known), location, and date. Mosses and liverworts may also be pressed, as described on pp52-3. They require less pressure, more changes of paper, and more time than wildflowers.

ALGAE, FUNGI, AND LICHENS

Algae, members of the Kingdom Protista (which also includes protozoans), are among the most primitive organisms that can be considered plants, though they are no longer classified in the plant kingdom. Most are photosynthetic. They are single-celled or multi-celled, and often occur as filaments or colonies. Many are mobile, usually moving by means of whiplike "tails" called *flagella*. Long strands of kelp and other seaweeds cast up on the beach, green pond scum, and greenish or brownish film on the side of an aquarium are just a few forms of algae.

Aside from their infinite value as the base of most freshwater and marine food chains, algae also have human uses. One of the most extensively used products of algae is algin, produced from giant kelps and other brown algae. Algin is employed in the manufacture of many food products, beer, pharmaceuticals, latex paint, paper, and textiles. Another algae-derived material with wide applications is diatomaceous earth. This powdery, abrasive substance is quarried from massive deposits of these aquatic organisms where past geological events raised them above the sea level. Its abrasiveness is due to the glassy shells of diatoms, a type of golden-brown algae, which are composed of up to 95% silica, an ingredient of glass. Diatomaceous earth is used extensively in filtration systems, as high-temperature insulation, and in the manufacture of metal polishes, toothpaste, light-reflecting paint, and prefabricated construction panels.

Above Many species of microscopic algae grow in large colonies. Here, long strands of filamentous algae were left by receding floodwaters of a river polluted by fertilizing agents.

Fungi

Though not classified as plants, members of the Kingdom Fungi are nevertheless plant-like in many respects. Unlike most plants, however, fungi are not photosynthetic, but absorb their food in solution from their surroundings. They are members of a general group of organisms known as *saprophytes*, which gain their nutrition by decomposing dead

Left Mushrooms, such as this species of *Amanita*, are the visible spore-bearing parts of fungi. One of the two major groups of decomposers in nature, fungi absorb their food in solution, which requires that the major portion of the fungi grows inside its host or underground.

What you can do

Identify fungi

Fungi are much more difficult to identify than wildflowers, which have characteristic leaf shapes and floral types, with consistent arrangements and numbers of floral parts. You cannot always compare fungi with illustrations to determine their identity, since many species have look-alikes, and no single field guide has pictures of them all. If, however, you make careful observations regarding the fungus and record them in a field notebook, this information can then be compared with a reference book to narrow the possibilities. Here is a standard form you can fill out and keep for each new fungus you find. Doing so will sharpen your powers of observation and build your knowledge of fungi.

Name (once identified) ______
Location ______ Date ______
Nearby vegetation ______
Substrate (soil, wood, etc.) ______
Growth pattern (single, cluster, etc.) ______
Stem: Shape ______
Size (length and diameter) ______
Texture ______
Shape of base ______
Annulus ______
Cap: Shape ______
Size ______
Color ______
Texture ______
Odor ______
Underside (gills or pores) ______
Gills: Color ______
Attachment to stem ______

Take spore prints

The color of spores and their arrangement in a spore print may provide a valuable clue to the identity of a mushroom. Permanent spore prints of mushrooms you've identified are useful additions to your field notes.

Cut the stem off a mature open mushroom and place the cap gill-side or pore-side down on a 3 × 5 index card (larger if necessary). You may need to use dark paper for light-colored spores. Cover with a bowl to exclude drafts and leave it for several hours or overnight. Afterwards, lift the bowl slowly and carefully remove the cap. The print will smudge easily, so to create a permanent record, spray with clear lacquer or artist's fixative, keeping the nozzle at least 18 inches from the print. Apply three coats and label the print with the name of the mushroom (if known), the location where found, and the date. Store your prints in envelopes for added protection.

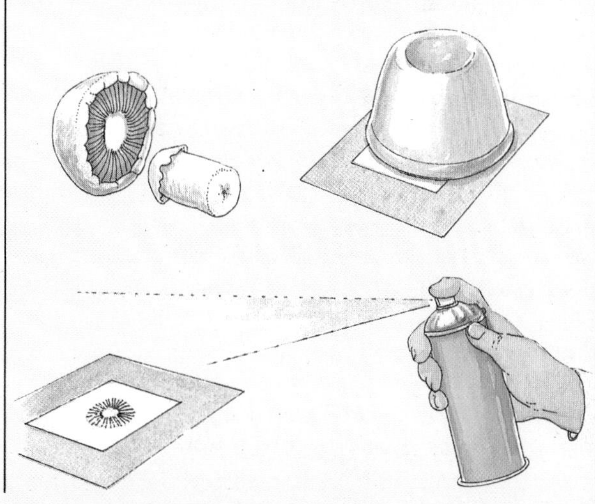

organic matter. Fungi, along with bacteria, are nature's recyclers, without which continued life on earth would be impossible.

Most species of fungi familiar to the practical botanist belong to the Class Basidiomycetes, the club fungi. These include mushrooms, puffballs, shelf or bracket fungi and stinkhorns. The structures often recognized as "fungi" are just the fruiting bodies of the organism protruding above the surface in order to release spores into the atmosphere. The bulk of the fungi, a mass of thread-like *hyphae* known as a *mycelium*, remains hidden from view, embedded in the material from which the fruiting body protrudes. The mycelium secretes enzymes into its surroundings in order to break the material down into soluble substances which can then be absorbed directly into the cells of the hyphae.

Mushrooms, also called toadstools, are composed of tightly interwoven hyphae, forming a stalk and an umbrella-like cap. The stalk may have a ring, or *annulus*, the remnant of a membrane that once extended from the stalk to the cap but tore as the cap opened. In some species, the underside of the cap consists of membranous *gills* radiating from the center. These gills are covered with

Below An emerging mushroom often leaves a ring of tissue, called the annulus, as the cap tears away from the stalk. Spores are produced in either gills or pores.

Right Spores fall from the gills or pores on the cap's underside, to be dispersed to suitable sites by wafting air currents.

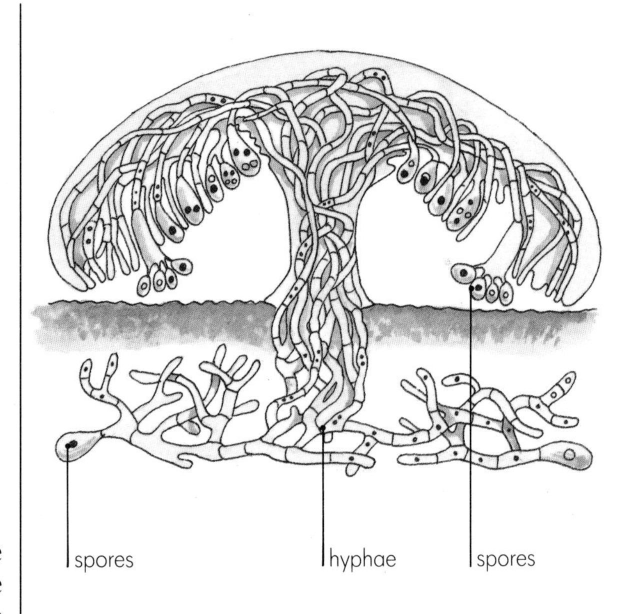

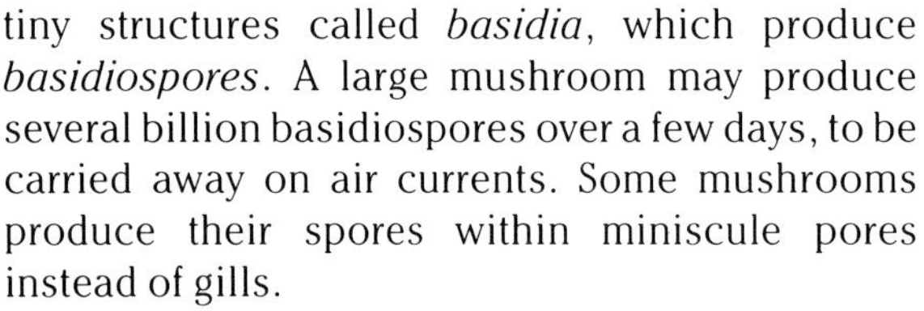

tiny structures called *basidia*, which produce *basidiospores*. A large mushroom may produce several billion basidiospores over a few days, to be carried away on air currents. Some mushrooms produce their spores within miniscule pores instead of gills.

Another familiar type of fungi, shelf or bracket fungi, grows horizontally on dead wood. Bracket fungi shed their spores through gills or pores. Puffballs, which have no stalk, produce their spores internally. When disturbed, the puffball is compressed, expelling air and spores through a pore in the top and through breaks in the outer covering.

Lichens

These are a prime example of mutualism, a type of symbiotic relationship, for lichens actually consist of two different organisms. An alga and a fungus are combined in a spongy body called a *thallus*, which ranges from a fraction of an inch up to several feet in diameter. The alga produces food for both organisms, and the fungus absorbs and retains water and minerals for both, and protects the alga from harmful light intensities. The fungus also secretes a chemical that accelerates photosynthesis in the alga, so it can feed both organisms.

Lichens are loosely grouped into three categories based on their growth forms. Crustose lichens frequently form brightly colored crusty patches, firmly attached to rocks and tree bark. Foliose lichens have leafy, often wrinkled, overlapping thalli, loosely attached to trees or rocks. The thalli of fruticose lichens are cylindrical and branched, often resembling miniature shrubs.

Their mass is attached to the ground or to vegetation only at one point, and sometimes hangs from branches. Although lichens are frequently attached to other plants, they do not parasitize them in any way.

Lichens are incredibly hardy, and grow under some of the most extreme conditions on earth. They can withstand prolonged periods of drought interspersed with wet spells. When dry, the upper layer of the thallus becomes opaque, effectively shutting down photosynthesis in the alga and plunging the lichen into dormancy. Lichens grow quite slowly, at a maximum rate of one-half inch per year. Some have been estimated to be 4,500 years old and are still growing. One condition they cannot tolerate, however, is air pollution, particularly sulfur dioxide. It is possible, in fact, to calculate the amount of sulfur dioxide in the air above a given area simply by mapping the occurrence or absence of certain lichen species.

Above The cup-shaped, spore-bearing structures of pixie cup lichen depend on a direct hit by a raindrop to dislodge their spores and disperse them in all directions.

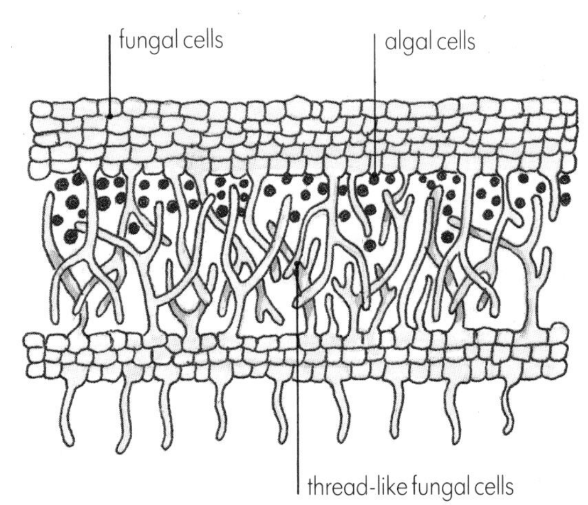

Left British soldiers, a lichen named for the red fruiting body, similar in color to the uniforms worn by British soldiers during the Revolutionary War, is common east of the Rocky Mountains.

Above Lichens are fine examples of mutualism, a relationship in which both organisms benefit. This is a cross-section of a typical foliose lichen, showing the algal cells underlying a layer of fungal cells which protect the algal cells and regulate the light reaching them.

ECOLOGICAL SUCCESSION

Ecological succession is the process by which vegetative communities develop. It is also the land's way of healing itself following a disturbance. All disturbances, whether logging, fire, glaciers or volcano eruptions, eventually come to an end. As soon as they do, the process of ecological succession begins. The stage at which it begins and the time needed to re-establish a stable community both depend on the severity of the disturbance.

Succession is an orderly transition from one plant community to another. The entire series of communities is called a *sere*, and each intermediate step is called a *seral stage*. In each seral stage except the final one, the dominant plants thrive, but at the same time alter the local growing conditions, making them more favorable for their successors than for themselves. As this occurs, *pioneer species*, the first plants of the next seral stage, begin to move in through their individual means of dispersal, further altering the habitat. The rate of succession slows with each stage, until a stable, self-maintaining stage, or *climax community*, is achieved. At this point, the area undergoes no more major changes until another disturbance takes place.

Primary succession begins in areas devoid of and unchanged by other organisms, such as bare rock exposed by glacial scouring or geological upheaval, new rock formed by volcanic eruptions or newly formed sand dunes. In these cases the initial stage is called a *xerosere*. Primary succession also occurs at the edge of a lake or pond as it gradually fills with sediment, and is then referred to as a *hydrosere*. Soil formation occurs on bare rock through the process of *weathering*, in which rock is broken into increasingly smaller pieces. Physical weathering is caused by forces such as alternate freezing and thawing of water in tiny fissures in the rock's surface. As water freezes, it expands, acting as a wedge to widen the crack and also forming new ones. When the ice thaws, water seeps farther into the cracks, and the process is repeated again.

Chemical weathering also aids soil formation. Lichens are typically the first plants to colonize a rock. Their acidic secretions slowly dissolve the rock, widening fissures, creating depressions, and liberating small particles. Chemical weathering also results from rainwater which may become mildly acidic because of atmospheric conditions or from draining over soluble materials. Lichens also help trap debris washed or blown over the rock, or debris may settle in cracks and depressions on its own, creating little pockets of sediment that retain moisture for short periods, encouraging the growth of small annual plants.

Soil, however, is more than pulverized rock. A very important component is *humus*, organic material derived from decaying plants and animals. Humus greatly increases the water retention of soil, and also provides a rich source of nutrients for plant growth. Each generation of plants adds more humus to the soil. In primary

Below A classic example of secondary ecological succession is the transformation of an abandoned field to a mature climax forest. Each seral stage, excluding the climax community, alters the habitat by increasing humus, water retention, and shade, thus creating conditions more suitable for plants of the succeeding stage than for the current inhabitants.

bare field   grassland 1-10 years   grass-shrub 10-25 years

Left Sumacs are common pioneers of the shrub-sapling stage of ecological succession. This staghorn sumac is abundant in the Northeast. Smooth sumac and shining sumac are more widespread, occurring throughout the eastern United States.

Right Red maple and gray birch are two common pioneer trees in abandoned fields. Red maple is found throughout the eastern half of the continent, while the range of gray birch is limited to the Northeast.

succession, the increase in soil depth is very gradual; it may take up to 10,000 years for an inch of topsoil to form. In the meantime, increasing soil depth and the accompanying increase in soil moisture and organic matter favors the germination of certain annual species, followed in time by perennial grasses, mosses, and herbaceous perennials, each making their own contribution to the soil buildup. Under the best of conditions, these will be followed by shrubs, small trees, and finally a mature forest.

Obviously, primary succession is a very long process. In contrast, *secondary succession*, which occurs following a disturbance to an area that already supported an advanced seral stage, is much quicker. In most cases, soil is already well established in the area. One familiar example is an abandoned farm field. Once the farmer ceases to work the field, grasses and annual plants quickly invade during the next growing season. Typically these are the so-called "weeds," exotic species that have adapted over thousands of years to take advantage of human environmental disruptions in Europe and Asia. Following the annuals are perennials, many of which are also exotics, then shrubs, shade-intolerant trees, and finally shade-tolerant trees form a climax forest.

There are many variations on the theme of succession, some of which are examined in greater depth later in this book. Generally, however, species composition changes continuously until a climax community is reached, but the change is usually much more rapid in the early stages of secondary succession. Also, the total number of species present increases in the earlier stages of succession, but tends to stabilize in the later stages. Finally, the end result is a climax community that is stable and self-maintaining.

pine forest 25-100 years

oak-hickory forest climax 100 plus years

BOTANY IN ACTION
THE TOOLS

The sage advice that one should use the right tool for the job applies to scientists as well as to carpenters and mechanics. You can study botany without tools, but you will advance much farther and faster by using a few basic ones. Some specialized items may have to be purchased through a biological supply house, but most of what you need can be had sooner and cheaper from retail stores. On the right is a basic list of those items a practical botanist will find most useful.

Field guides

Field guides providing clear illustrations, which include characteristic identifying features, are invaluable in identifying and learning about new plants. Guides also limit the number of possibilities by listing the plant's likely geographic range and habitats. Finally, they give a clear physical description of the plant, and include an illustrated glossary to explain unfamiliar terms. Any field guide you purchase should have these features. There are many good field guides, although none can cover North American botany in its entirety.

Unfortunately, learning often ceases once a plant is identified, but it doesn't have to be so. After identifying a plant, read the text for that species again. Field guides often contain fascinating details about folklore, medicinal uses, historical associations and edible parts of the plant. Also, visit your local library regularly and look up your newest botanical acquaintances. Knowing plants' cultural and natural histories reveals the many ways in which plants interact with the rest of their world.

Keys

Identification keys, sometimes known as dichotomous keys, are excellent companions to field guides, but harder to obtain. "Dichotomous" means divided or dividing into two parts. At each step, the key presents two, or sometimes more, characteristics, only one of which will apply to the organism under scrutiny. Depending on the option selected, you are directed to another section of the key to repeat the process until, finally, the plant's or animal's identity is revealed. Because keys only reveal the identity, you should follow this up by reading about the plant in a field guide.

Above Copious note-taking in the field is an invaluable practice. Personal observations stick with you much longer than anything you read.

Keys are normally available only for higher plants, such as trees, wildflowers, and ferns. They are sometimes found in bookstores, but you can often obtain them from the botany department of your local university, from university bookstores, or from a nature center. The "Finder" series is a practical set of keys in booklet form available in many bookstores, or from Nature Study Guild Publishers, Box 972, Berkeley, California, 94701.

Field notebook

A medium-sized, unlined notebook, about 4 × 7 inches, is a good compromise between convenience and size. It should be strongly bound, and have high-quality paper that will not tear easily. Ziploc bags make tough, waterproof storage cases for your notebook in the field. (Use them for maps, field guides, and identification keys as well.)

Tools of the practical botanist

Identification
Field guides
Identification keys
Examination
Hand lens
Probe
Tweezers
Small plastic tray
Penlight
X-acto knife
Recording
Field notebook
Pen
Pencils
Tape measure
Ruler
Camera and accessories
Collecting
Pruning shears (clippers)
Tree wound dressing
Large rigid plastic container
Absorbent paper
Plant press
Manila tags
Wax pencil
Trowel
Camel hair artist's brush
Small envelopes (for collecting seeds)
Miscellaneous
Day pack
Topographic maps
Compass
Swiss army knife

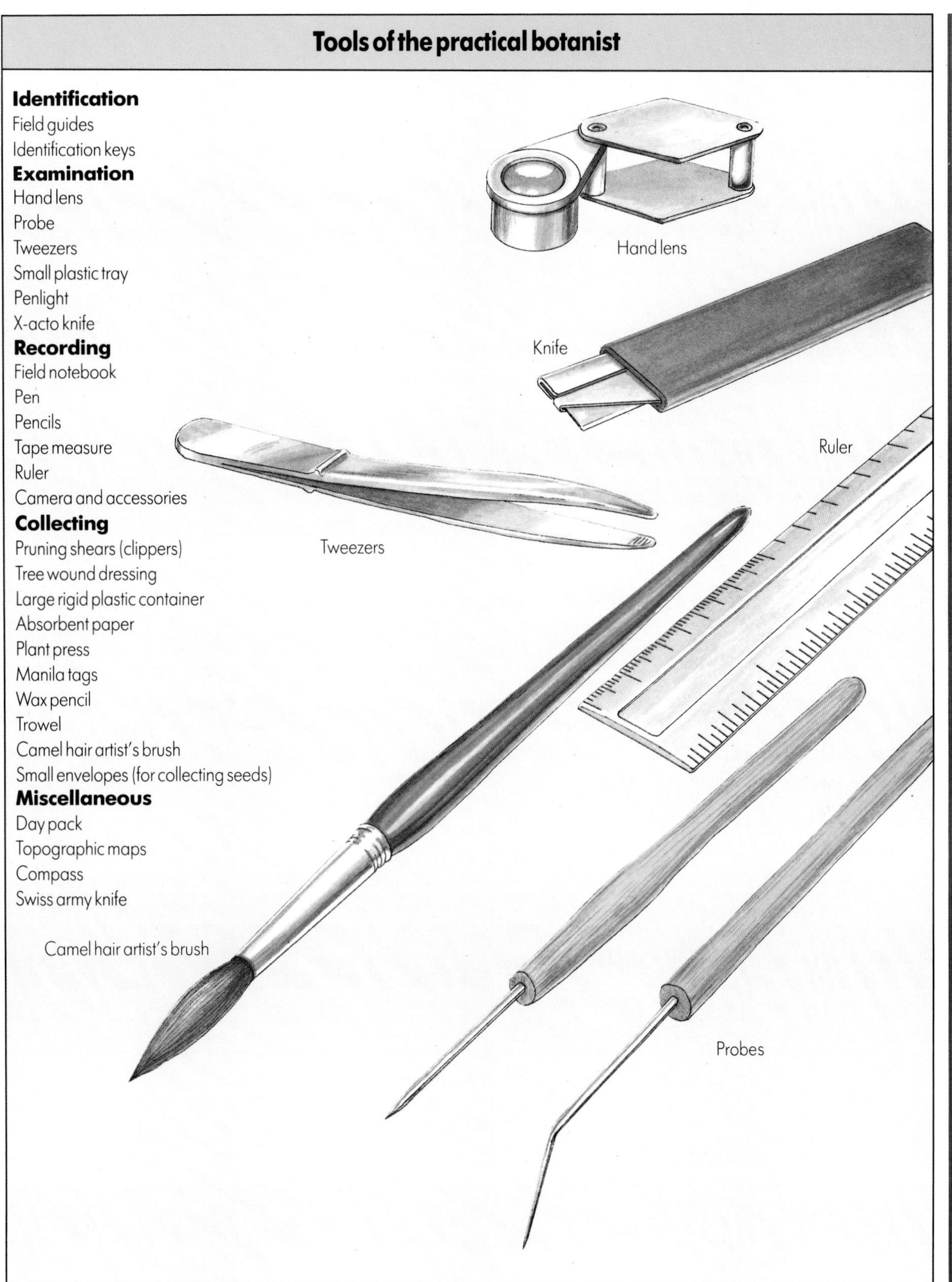

Photography

Probably the best way for practical botanists to "collect" specimens is to photograph them. It is faster, simpler, and, most importantly, more environmentally sound than removing plants from their habitats and preserving them. Photography is not the daunting endeavor it once was, especially with today's auto-everything cameras. The 35mm single-lens reflex models are the most versatile and have the most accessories (see pp46-51).

Collecting

When collecting wild plants, you must use good judgment and follow ethical guidelines. Humankind has long since passed the point where we can do as we wish without regard for environmental consequences, and removing even part of a plant may destroy it or interfere with its reproduction. Obviously no ecological Armageddon will befall us for simply removing a few daisies and butter-cups from a meadow. However, the rules listed below ensure that your actions will have minimal consequences.

To carry specimens home, use a rigid plastic container, such as those sold in the housewares department of supermarkets or department stores. The large, rectangular, airtight models work best. Layer specimens between sheets of paper, and tie a manila tag to each one with the collection date and specimen number written in wax pencil, which remains legible even when wet. Record the specimen number and date in your field notebook, together with the plant's name, if known, plus location, soil type (moist, dry, sandy, loamy, rocky, etc.), and other pertinent information.

Guidelines for responsible collecting

- Unless you are building a collection for serious study, take only the most common species.
- Collect only plants you can positively identify in the field. It is your responsibility to avoid taking threatened or endangered species. Your state or provincial conservation department can provide lists of the endangered species in your particular area.
- Do not remove the whole plant if part will suffice.
- When removing part of a plant, use sharp pruning shears to minimize trauma. Seal wounds on woody plants with tree wound dressing, available at hardware stores and garden shops.
- If you must remove an entire plant, replace the soil as well as you can.
- If you collect seeds, take only a few from one or two plants.

Hand lens

A good-quality hand lens is indispensable to the practical botanist. It aids identification and reveals minute details that help you understand the function and admire the beauty of structures that you might otherwise overlook. They are usually available from a biological supply house, though they are occasionally sold elsewhere. A magnification of 8× to 10× is optimal, but some models have several glass elements that can be used independently or combined to obtain varying magnifications. To view an object with a hand lens, hold the lens about two inches from your eye, then slowly bring the object into focus. Hand lenses are invariably dropped in the field, so tie brightly colored yarn to them so you can find them again.

Dissecting instruments

These are useful in separating minute structures for observation and for sketching. Ordinary, narrow-jaws tweezers from a pharmacy work well. Probes are available from biological supply houses, but you can make one by taping a stout needle to a six-inch section of quarter-inch wooden dowel, sold in hardware stores. A small plastic tray is useful for examining and dissecting specimens. In low-light situations, a small pocket flashlight is valuable for examining details. An X-acto hobby knife makes clean cuts for viewing internal structures. Use an old toothbrush holder to transport the probe and X-acto knife safely.

Right Hold the lens about two inches from your eye and bring the specimen to about the same distance from the lens.

STARTING AROUND HOME

One of the best ways to fuel your new-found interest in botany is to examine the botanical goings-on around your home: the "weeds" you ignored for all of these years, the trees whose identities you always meant to learn but didn't, and the mosses and lichens you just plain didn't see before.

Taking inventory

Some educators deplore listing species as counter-productive to learning, assuming that interest wanes once you attach a name to an organism, but the opposite can be true. Each new addition to the list of plants around home reinforces your interest, and just when you think that you've listed every plant, what a joy it is to discover a new species! You can then investigate how it got there, when, and why; whether it is an annual, biennial, or perennial; or even find a possible parent plant by venturing a little farther afield. If it is a flowering plant, you can discover when it will flower and for how long, what the fruit looks like, and what insects, birds, or mammals it might attract. These questions and many others may come to mind with each new discovery you make.

Where to look?

An abandoned vegetable garden is a prime location for exotic wildflowers seeking disturbed soil in which to quickly put down roots before their competitors do. The waste soil along the roadside is another favorite habitat of theirs. A well maintained flower garden is also a challenge, but since you planted the flowers and presumably know their names, the challenge is not to identify them but to place them in their proper families. This can be formidable when dealing with selectively bred domestics, but the family descriptions in a wildflower guide may help. Trees are obvious subjects. Interesting moss is often found in the protected niches between exposed tree roots. Moist, shady, undisturbed places harbor moss, and maybe ferns or liverworts. Lichens colonize tree bark and exposed rocks and bricks, but they are slow growers and intolerant of air pollution, so they may be entirely absent. Even a puddle under a rain spout is likely to host a thriving colony of algae.

Ironically, the largest expanse of vegetation likely to be around your home, the lawn, is also

Above Your own yard or a nearby city park both offer more botanizing opportunities than you might think. Even the ubiquitous dandelion proves a fascinating specimen.

Above Many seasonal changes occur in autumn as plants prepare for winter. Here the brilliant fall foliage of quaking aspens decorates a mountain road in California.

botanically monotonous. It is a monoculture, an artificial habitat composed primarily of one dominant plant type that is expensive to maintain in terms of time and energy. Monocultures are unstable, and invariably diversify once the forces maintaining them are eliminated. To demonstrate this, leave a small patch of lawn undisturbed for a season and observe. Not only will you notice that the grass grows tall and flowers, sometimes confirming the existence of more than one species of grass, but that members of other families become evident as well. Some, like common dandelion, bluets, and common plantain, may have been there all along but were held in check by the lawn mower. Others are pioneer species, able to establish themselves only in the absence of herbicides and the motorized whirling dervish.

When to look

Any time is the right time. Spring reveals new growth, buds open, leaves unfurl, flowers blossom, and shoots elongate. Summer is the season of maturity, when late bloomers flower and the fruits of early bloomers begin to ripen. As seeds develop, examine them and try to guess their method of dispersal. In the fall, plants begin preparing for their dormant period, and deciduous leaves change color. You can record when the first leaves begin to change, whether the same plants change color first every year, whether the foliage of one plant turns the same color every year, and what conditions produce the most brilliant colors. In winter, you can focus on woody plants and evergreens, but the dried stems and seed heads of last season's herbs also beckon, as do lichens on rocks and bracket fungi on dead trees.

It's fascinating to observe the same plants on a daily basis and note the changes in your field notebook; referring to these notes in future seasons will remind you of upcoming events and you can then look in the right place at the right time to witness them.

Your field notebook

Field guides and identification keys aid identification and provide excellent background information, but *you will learn more about plants, and retain more of what you learn, from your own detailed observations than you will from reading about them!* Notes on anything you find interesting or significant serve you for the rest of your life. Basic information on each page should include the date, departure and return times, weather conditions, temperature, and locations and habitats visited.

Always record observations on the spot. If you trust them to memory, important details are lost. Even a short note with a few key words will jog your memory, and you can expand upon it later. Always include sketches with your notes. However crude, sketches are great reminders of exact details, such as leaf shape and arrangement, sizes, and flower structure.

A field notebook's value is enhanced by keeping a loose-leaf diary or logbook at home, into which you can transfer your notes, reorganize and expand them, and co-relate them with your readings and past observations. Do this religiously after each excursion, and you will amass a body of knowledge that is unique; some of it will not be found in field guides, and may never have been recorded before.

Don't expect to learn everything about a particular species on your first encounter. Think of each one as a jigsaw puzzle with the pieces askew; with each rendezvous, you can put a few more pieces in place. Gradually, over weeks, months, or even years, a picture begins to emerge. The background of the picture, too, becomes more distinct as you learn more about how the plant interacts with its surroundings. Keeping a field notebook allows you to do this; without it, many of the puzzle pieces are displaced before you chance upon that species again.

Cataloguing

As well as a notebook and in place of a logbook, you may find it useful to start an index card file on the species you come across. Use one card per discovery and record its name, date found, relevant information, and the page number of your notebook on which you refer to it. Number each card and use the corresponding number to identify photographs, specimens, and notebook entries. It is easier to have notes on file this way than to search through notebooks for information.

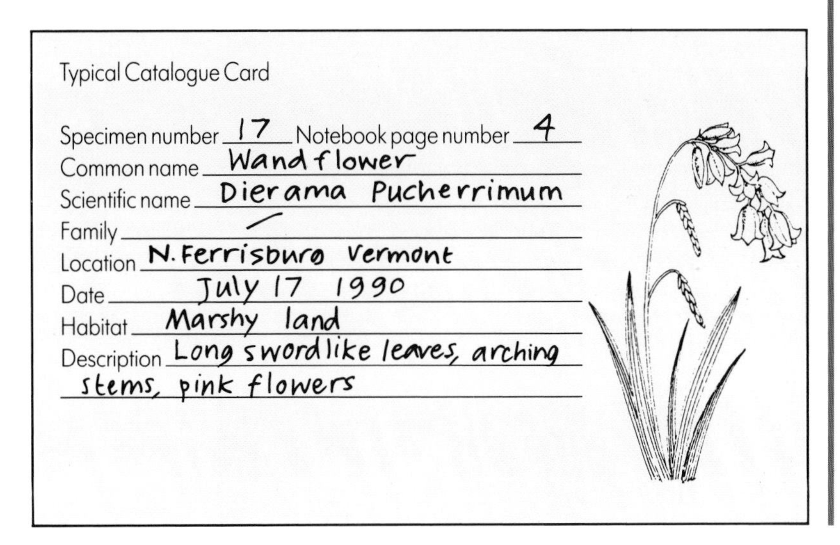

Typical Catalogue Card

Specimen number 17 Notebook page number 4
Common name Wandflower
Scientific name Dierama Pucherrimum
Family /
Location N. Ferrisburg Vermont
Date July 17 1990
Habitat Marshy land
Description Long swordlike leaves, arching stems, pink flowers

Above and **right** The combination of a meticulously-kept field notebook and a detailed index card file is invaluable to the botanist.

SKETCHING

Sketches should become an integral part of your field notebook, since your mind relates to pictures better than it does to words. Even the crudest sketch takes you back to the time you made it, and your mind will fill in the details. Drawing skills, luckily, improve with practice. The value of a sketch is not measured aesthetically; its value lies in what you remember from it.

Sketching trains you in precise observation, a skill critical to any naturalist. If you overlook details while drawing from a model, you will realize quickly that something is absent and, almost subconsciously, begin searching the drawing for the missing features. Again, quality is not important, but attention to detail is. A sketch does not have to include the entire plant. Often, detailed illustrations of different parts of the plant, together with a rough diagram to show how the parts are arranged, are more instructional than a rendition of the whole organism, especially if it is large.

Before drawing a plant, try to answer five basic questions about its construction.

What are its roots like?
Some plants have fibrous roots, some have taproots, and some have both. Adventitious roots may be present and some plants have no true roots at all. Some plants have individual root systems, while others share a spreading root system. Anchor each plant in your drawings in the way it grows in nature. Draw roots from the top down to portray their tapering and probing nature.

How do its stems grow?
Stems can be straight or branched, long or short, erect, arching, or trailing. Some plants have stout stems, others are slender and willowy. Depict these qualities in your drawings, recording, too, the shape and location of buds. Draw stems from their base upward to capture their tapering shape.

Does it have leaves, and if so, what are they like?
Look at their shape, margins, texture, position, and vein pattern. Leaf arrangement may be alternate, opposite, whorled, or basal.

How are flowers, fruits, or seed pods attached?
Note whether they are solitary or grow in clusters. Observe their location on the stem. Examine the flowers with your hand lens and move their parts around with a probe in order to see their exact shapes and arrangement.

Are there any obvious tropisms?
Plants move as surely as animals do; they just move differently and for different reasons. Observe these movements and try to convey the impression of

Right To achieve an accurate representation, study a plant thoroughly before sketching and try to discern the direction in which each part moves or grows. Start at the base of that part and sketch the lines running in the same direction.

Left True to the maxim, "a picture is worth a thousand words" sketching sharpens your powers of observation and helps you recall important details later. A useful sketch can be rough, or well finished.

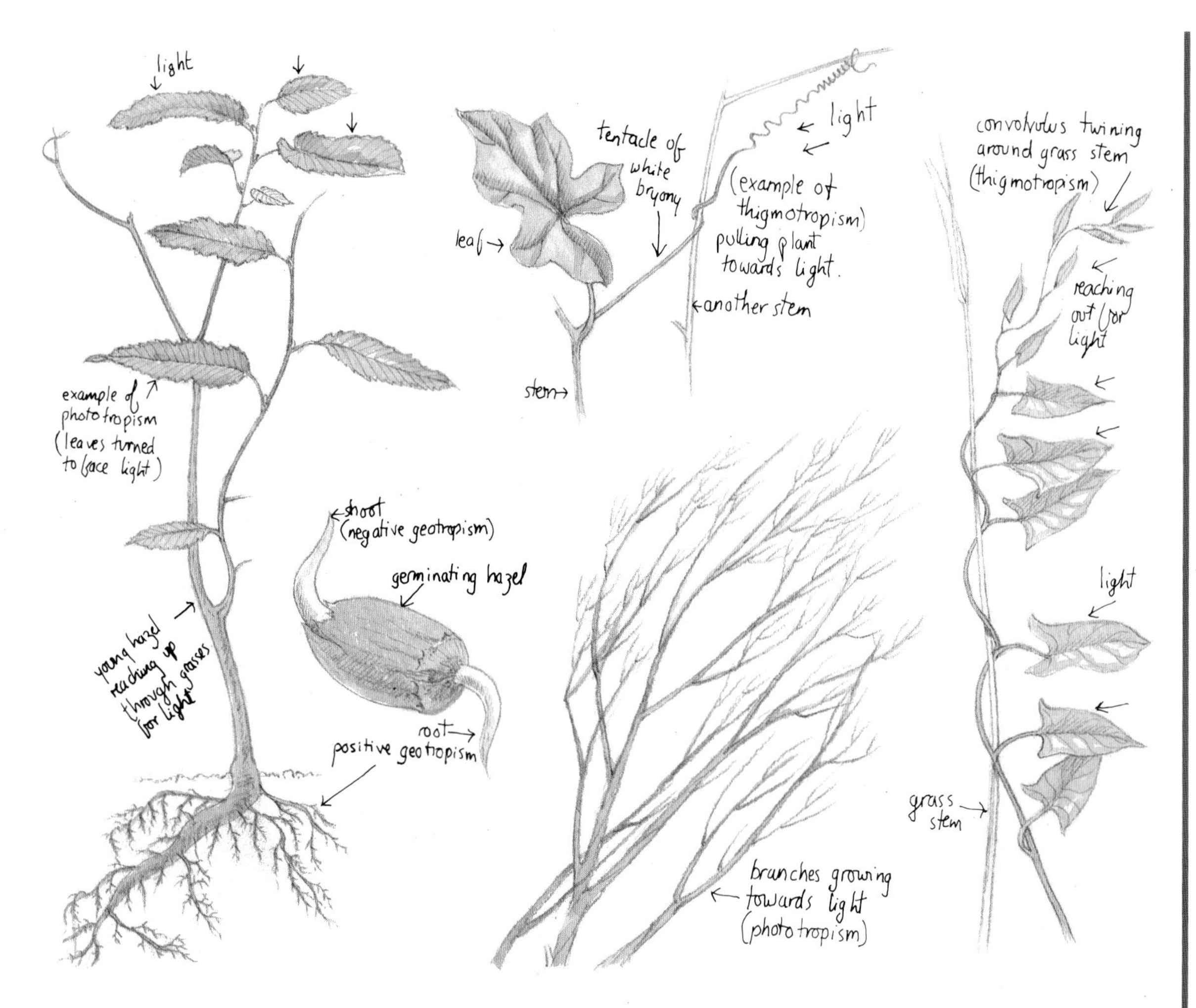

them in your sketches. Growing plants move when stimulated by light, gravity, or touch; such movements are called tropisms.

Geotropisms are movements stimulated by gravity. A root growing down into the soil is an example of positive geotropism. New shoots reaching up out of the soil demonstrate negative geotropism. To illustrate these, use downward strokes of your pencil when drawing roots, upward strokes when drawing sprouts.

Once above ground, plant stems and leaves move in response to the presence or absence of light. Such movements are called phototropisms. Herbaceous plants and young woody plants in semi-shaded areas move every which way to reach light, both stems and leaves bending toward the light source. To show this, draw the stem in the direction of the light source, then sketch the leaves at right angles to the light. Crowded trees grow straight and tall, reaching for a sunny spot in the forest canopy. Use upward strokes to depict the effect of reaching for light. Conversely, in open areas, a plant's stems and leaves tend to spread out evenly in all directions.

Plant movements in response to touch are called thigmotropisms: tendrils of bean plants and grape vines, for example. Draw free tendrils with a random, wandering line, as if they were feeling for an anchor. Once they touch a stationary object, tendrils wrap tightly around it, which can be depicted with a tight spiraling line. The twining climb of some vines, best illustrated with a weaving or spiraling line, is due to the stem's bending in response to the touch of its support while also bending toward the light and away from gravity.

PRACTICAL PHOTOGRAPHY

For practical botanists, there is only one group of cameras to seriously consider; the 35mm single-lens reflex (SLR) models *with interchangeable lenses* offer the best compromise between quality, versatility, convenience, ease of operation, and cost. They are the overwhelming choice of nature photographers, including top professionals, around the world.

The quality of photographs taken by a competent photographer using a 35mm SLR camera is difficult to beat. In terms of versatility, 35mm SLRs are second to none, and possible subjects range from the microscopic to the astronomical. Simply by changing lenses, you can go from shooting close-ups of a buttercup to taking telephoto shots of a skittish songbird in a matter of seconds. They are also light enough to go anywhere. Ease of operation varies, but the newest auto-focus, auto-exposure cameras are literally as simple as point-and-shoot, although this by itself does not guarantee a good photograph. Finally, the cost of 35mm cameras and basic accessories is well within most budgets. Second-hand cameras in good condition can often be found in camera shops or through classified ads, but always have any second-hand equipment examined by a qualified camera repairman before purchasing it.

The choice of brand is largely a matter of personal preference. For background information, write to the major photography magazines and request their field test reports on models you are considering. These magazines also publish annual buyer's guides, which show what's available. Note the quality and variety of lenses and accessories available for different brands, but you needn't buy top-of-the-line equipment to achieve quality results. Remember the story of the photographer and the writer who met at a party. "I saw your exhibit at the gallery last week," said the writer. "You must have a very good camera!" To which the photographer replied, "Why, thank you. By the way, I really enjoyed your latest novel. You must have a very good typewriter!"

Certain accessories are quite useful to the practical botanist, others are less so. Lenses are labeled by their focal length, which in turn determines their perspective. Most cameras come with a standard 50mm lens, though you can purchase the camera body alone and select your own lenses. A 50mm lens gives roughly the same perspective, or angle of view, as the human eye.

Wide-angle lenses have a focal length of less than 50mm. These provide a wider field of view than seen by the human eye, and consequently subjects appear smaller, although the difference is negligible until you get down to a focal length of 35mm or less. Wide-angle lenses are useful for photographing landscapes, habitats, and large subjects at a close distance.

At the other extreme are telephoto lenses, with focal lengths from 50mm to 1,000mm or more. Only medium-length telephoto lenses, up to approximately 200mm, have any practical application in botany. Good quality longer lenses are

Features to look for when selecting a camera

Mandatory	Optional
Single-lens reflex	Auto-exposure with manual override
Interchangeable lenses	Auto-focus with manual override
Depth-of-field preview	Motor drive
Built-in light meter	Dedicated TTL flash capability
Self-timer	Auto-rewind
	Lock-up mirror
	Interchangeable focusing screens

prohibitively expensive, heavy to carry and nearly impossible to hold still enough for a clear image without using a tripod. Telephoto lenses provide a narrower field of view, and subjects appear larger than when seen by the unaided eye. They are used primarily when you cannot get as close to your subject as you wish; or in order to isolate it and remove a cluttered foreground.

High-quality zoom lenses, providing variable focal lengths, are extremely practical for field work. The entire range of focal lengths from about 24mm up to 210mm can be covered with just two lenses, and in some cases even one lens suffices. This leaves you relatively unencumbered and able to carry other gear. By adjusting the focal length, you can compose a photograph exactly as you want it without having to move the camera closer to or farther from the subject.

Close-up equipment is highly useful in botanical photography, revealing a whole world of minute and fascinating detail. Macro-lenses can be used at their designated focal length, but you can also continuously focus down to close distances, usually a few inches. They do double duty, and many photographers prefer to purchase a 55mm macro-lens instead of the standard 50mm lens offered with the camera body. Macro-lenses come in a variety of focal lengths, the major difference

Above A macro lens can focus closer than a standard lens of the same focal length, so that the image of the subject on the film is larger. The same effect can be achieved by mounting extension tubes or a bellows between any standard lens and the camera body.

Right These two photographs, taken at the same distance from the subject, illustrate the difference in perspective between a 50mm lens, left, and a wide-angle lens (24mm in this case) at right. As the focal length decreases, image size decreases and range of focus, called depth-of-field, increases at any given aperture, and vice versa.

being how close you must be to the subject to achieve maximum magnification. The best macro-lenses give 1:1 reproduction; the image on the frame of film is the same size as the actual subject. The same results possible with macro-lenses can be achieved with standard lenses by adding dead air space between the camera body and the lens, usually by means of extension tubes or bellows. Extension tubes are much less expensive and more practical to use, but offer somewhat less versatility in achieving a specific image size. Regardless of the close-up equipment used, your image quality will be greatly improved with the use of a tripod.

Film types

Film types are nearly as varied as camera models. What's right for you depends on how you want to use your images, the conditions under which you are shooting, and budget. Print films are developed into negatives, from which a print is made on light-sensitive photographic paper. Prints are easy to display and view, either in an album, or framed on a wall. Black-and-white prints produce an accurate representation of form and texture, but with such a rich spectrum of colors in the plant kingdom, you cannot achieve a complete portrayal of a plant in black and white. Color prints are more aesthetically pleasing and more accurately portray natural subjects than do black-and-white prints, but they are also the most expensive film to develop, thus limiting botanists on a tight budget.

Color-slide films, also known as color reversal films, are the workhorse of professional nature photographers, especially those shooting for publication, since transparencies (and black-and-white, to a lesser extent) are used almost exclusively in publishing. Transparencies cannot be viewed easily without a projector, slide viewer, or a hand lens and light table. Their major application, aside from publishing, is in illustrating lectures. Slide film is much less expensive to develop than color print film, a boon to the botanist on a budget. You can also have prints made from slides at a custom photo lab, and although it costs more per picture than print film, you can afford to experiment with different exposures and compositions and print only those you like.

The other major variable associated with films is light sensitivity, better known as film speed and denoted by an ASA (American Standards Association) number. The higher the ASA number, the faster, or more light sensitive, is the film. Faster films require a faster shutter speed *or* a smaller lens opening under a given light intensity than do slower films. Greater light sensitivity is achieved by making the grains on the film emulsion larger, consequently photographs produced with faster films are more grainy than those produced with slower films; the lower the ASA number, the sharper the image. For this reason, publishers generally will not consider using images produced on film rated higher than 100 ASA. High-speed films are most useful under low light conditions without a flash, or when you want to use a very fast shutter speed to freeze movement.

Above Slower films, designated by a low ASA number (which range from 25 to 1000) are less sensitive to light than fast film, but yield photographs of superior clarity, highly favored for publication. This sheep laurel was photographed with 25 ASA Kodachrome.

Above Faster films are useful in low light conditions, where electronic flash is impossible or undesirable. The large grains in the film emulsion that enhance light sensitivity also result in grainy photos, as this image of a mushroom, photographed with 400 ASA Ektachrome, shows.

Exposure

Whether you use natural light or electronic flash depends on the effects you wish to obtain, the amount of natural light available, and on how much your subject is moving. A photograph records light. To achieve proper exposure, the correct amount of light must reach the film. This is a function of the combined shutter speed and size of the lens opening, called the aperture or f-stop. Modern 35mm cameras have built-in light meters that measure the amount of light reaching the film. Most cameras can use this information to calculate the correct aperture for a given shutter speed, or vice versa; some will calculate both.

Getting the correct exposure is like trying to fill a bucket exactly to the top with water from a faucet. A slow trickle takes longer than opening the faucet all the way. If too much water is added, the bucket overflows; if too little, it is not as full as you wish. Correspondingly, correct exposure requires the use of a longer shutter speed with a small aperture than with a large one, and vice versa. Too much light results in overexposure with washed-out highlights, and too little light results in an underexposed image.

Whether or not to use electronic flash depends on several factors. For a natural-looking photograph, try to use the natural light available. Bright overhead sunlight creates harsh, unappealing shadows, but on smaller subjects, shadows can be softened by using reflectors made of cardboard covered with crinkled aluminum foil. Shadows may also be softened by constructing a light-diffusing lean-to from translucent sheets of plastic, such as those used to cover fluorescent light fixtures. Low natural light requires a large aperture, a slow shutter speed, or

Below This side-flowered miterwort was photographed using natural light, right, and electronic flash, left. An electronic flash lets you use a smaller aperture, thus increasing the range of crisp focus, or depth-of-field. The smaller aperture, in turn, darkens the background and isolates the subject, often a desirable effect.

Above These photographs of puffballs illustrate the effects of natural light, left, versus electronic flash, and the resulting greater depth-of-field, right.

both. Slower shutter speeds require a tripod to attain sharp focus, but any movement of your subject effectively prohibits slow shutter speeds, even with a tripod. In these cases, or to darken the background and emphasize the subject, electronic flash is the best method. Flash also fills in harsh shadows, freezes movement, and lets you use a smaller aperture to increase depth-of-field.

Focus

Another function of aperture size is depth-of-field, the area in focus within a certain range of distances from the camera. Using a given lens, a smaller aperture yields a greater depth-of-field than a larger aperture. To emphasize a subject by focusing on it but not the background or foreground, use a larger aperture and a faster shutter speed. For more depth-of-field, to portray a plant in its natural habitat, for instance, use a small aperture, and a correspondingly slow shutter speed. Focal length of a lens also affects the depth-of-field. Shorter focal lengths yield greater depth-of-field than longer focal lengths. Most lenses have a scale showing your depth-of-field in feet and meters from the

Recommended photographic equipment for the practical botanist

Option 1	Option 2
35mm SLR camera body	35mm SLR camera body
28mm lens	28-70mm zoom lens
55mm macro-lens	70-210mm zoom lens
105mm macro-lens	Extension tubes
200mm lens	Warming filters
Extension tubes	Polarizing filters
Warming filters	Electronic flash
Polarizing filters	Tripod
Electronic flash	Cable release
Tripod	Reflector
Cable release	Diffuser
Reflector	
Diffuser	

Above Depth-of-field, the zone of sharp focus in your photographs, is an important consideration in composition and in capturing the details you wish to record. These two photographs of the same day lily illustrate the effects of using a small aperture (f32) to achieve maximum depth of field (**left**) and a large aperture (f2.5) which yields a shallow depth of field (**right**).

camera, so you can calculate exactly which parts of your photographs will be in focus.

A sturdy tripod is the most effective accessory for improving the quality of your photographs, since most blurred images result not from poor focusing, but from camera movement. Although a tripod can improve sharpness at any shutter speed, the universal rule of thumb is not to hand-hold a camera when shooting at less than the reciprocal of the focal length of the lens. Put simply, when using a 50mm lens, you should use a tripod or other steady support if your shutter speed is less than 1/50 of a second. Hand-holding a 200mm lens requires a shutter speed of at least 1/200 of a second, and so on.

Guidelines for purchasing a tripod

- Buy the sturdiest, not fanciest, tripod you can afford. Before you buy, set up the tripod and lock everything in place, then wiggle, twist, and turn every part of it. Nothing should move, not even one millimeter.
- Select a tripod that lets you shoot at or near ground level, so you can photograph low-growing plants. Avoid tripods with reversible center columns, which let you shoot at ground level, but have a tripod leg between you and an upside-down camera – nearly impossible for most of us to do.
- Buy a tripod that extends to eye level *without* the center column extended, otherwise your tripod is converted into a much less stable, three-legged monopod.
- Make sure the tripod and camera are light enough to carry without undue strain. The strongest tripod in the world is completely useless if left at home because it is too heavy.
- Buy and use a cable release with your tripod, to prevent camera movement caused by depressing the shutter button with your finger.

PRESERVING PLANTS

The most practical and environmentally sound method of collecting plants is photography. You can "acquire" new specimens without disturbing them and without spending time and effort preparing and preserving them properly. However, if you want to establish a collection of plants, there are several methods of preservation.

Herbaria

A herbarium (plural, herbaria) is a library of dried and pressed plants organized into families, with genera and species arranged alphabetically within each family. This arrangement allows them to be easily located for reference, just as one would look up a book in a library. Properly preserved and maintained specimens have remained usable for more than 300 years, and herbaria are standard botanical references of universities and museums. Creating your own herbarium is fairly easy and involves making or buying a simple plant press and following a few elementary procedures.

A plant press consists of heavy blotting paper and cardboard sandwiched between two sheets of 12 × 18 inch plywood. It is usually held together by a pair of nylon webbing compression straps, but you can use 14 × 20 inch plywood and secure it with bolts and wingnuts at the corners. The dimensions should be slightly larger than standard herbarium sheets.

Proper handling and storage of specimens in the field is vital for successful preservation. Transport specimens in a rigid holder such as a large plastic food storage container, to avoid crushing. Keep them as moist as possible, but provide some ventilation. If you cannot press them immediately, gently wash the plants to remove any remaining soil and place them in a vase with fresh water.

When ready to press the plants, blot off excess water and arrange each one on an open sheet of newspaper so that the leaves and flower parts are not folded or overlapping. *Remember to keep the*

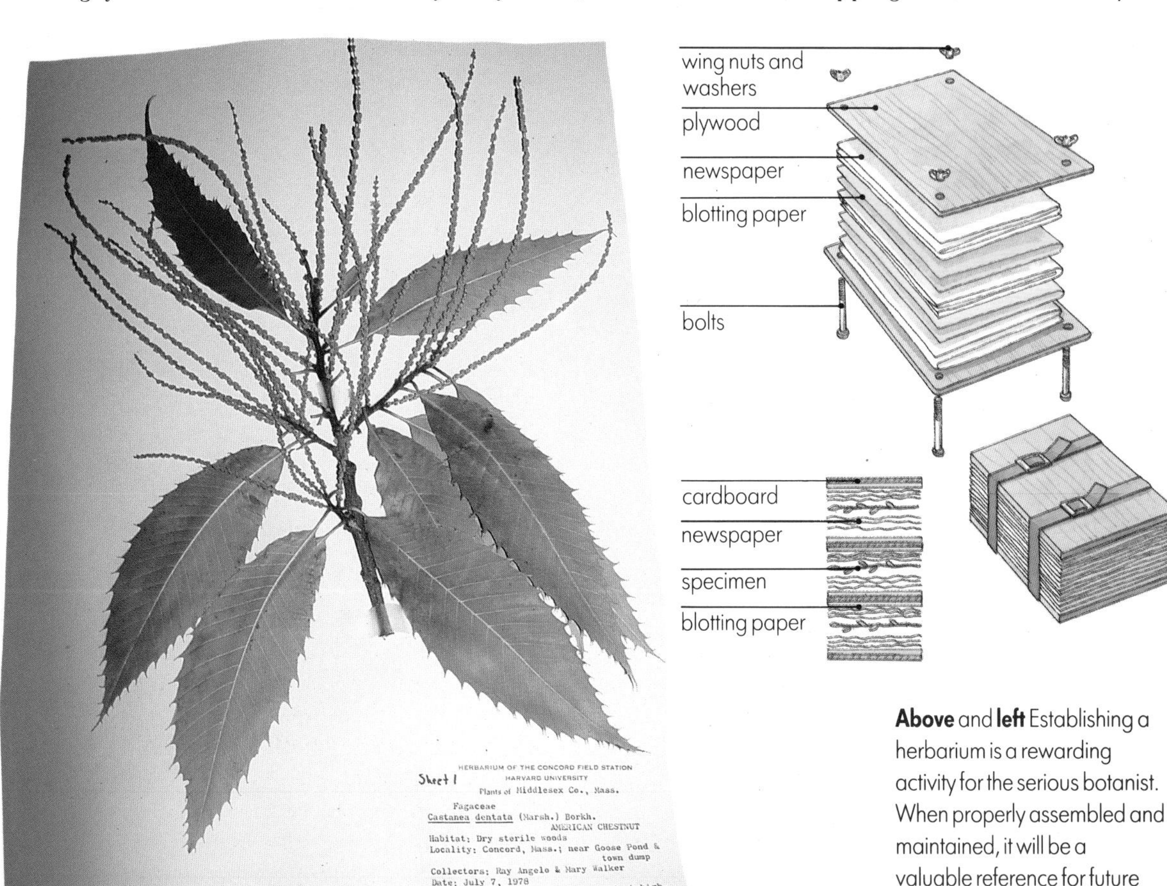

Above and **left** Establishing a herbarium is a rewarding activity for the serious botanist. When properly assembled and maintained, it will be a valuable reference for future generations.

identification tag attached that corresponds to the information recorded in your field notebook. Fold the newspaper over, place it between two blotters, and sandwich this between two sheets of corrugated cardboard. The air spaces in the cardboard allow ventilation to aid drying. No more than ten items should be pressed at a time. The wingnuts should be finger-tight. Keep the press and plants in a warm, dry room for two to four weeks.

The faster specimens dry, the better the color retention. Change papers every two or three days, replacing damp paper with fresh and tightening the press. (The blotting paper can be reused after it dries.) Large, fleshy plants require more paper and more frequent changes than those with a lower water content. You can use a microwave oven for super-fast drying, but you need a small press with *no metal parts.* Drying times vary, but two minutes on medium power is a good starting point.

When pressing bulky items, such as taproots and fruits, pack wads of paper around them to prevent crushing; larger fruits may have to be cut in half. Arrange aquatic and weak terrestrial plants for pressing by floating them in a tray of water, sliding a stiff piece of cardboard or parchment underneath them, and lifting them out of the water. (Soak saltwater species in fresh water overnight to remove the salt.) Allow the plant to air dry for about an hour before pressing it lightly in the plant press with extra sheets of paper. With large plants select key parts and press them in the normal fashion.

Try to use 100% rag-content paper for mounting dry specimens, since pulp-content paper deteriorates with age. The standard paper size for herbaria is 11.5 × 16.5 inches. Use good-quality, white library glue at strategic locations to mount the specimen, leaving the lower right-hand corner free for the identification card, which should also be 100% rag-content paper. Transfer all relevant collection information from your field notebook to this card, including common and scientific names, collection date, and geographic location, habitat type and soil type where found. Include your name as collector, in case you should ever wish to donate your work to a museum or university, and assign an identification number to each one for your cataloguing system. Place the mounted specimens in the freezer for three days to kill any pests, and use mothballs in the storage area to prevent insect damage. Encase each specimen in a clear plastic bag and store in manila envelopes.

Above Creating dried flower arrangements is an enjoyable hobby for the botanist with an artistic streak. Use cultivated varieties or the *most* common, non-native wildflowers.

Three-dimensional preservation

Three-dimensional specimens present some storage problems, but you may want to preserve some plants in this fashion, for study or just decoration. Each of the methods given below has its own advantages and drawbacks.

Air drying

Air drying is the simplest method of preservation, but air-dried flowers may not retain their shape and color as well as those preserved by some other methods. Air dry plants in a warm, dry, dark, well ventilated place, but not near a heat source, which causes increased brittleness. Check the humidity level with a hygrometer, available in most hardware stores; the humidity should be less than 50%. Leave space for air to circulate around the material; avoid overcrowding.

Many species dry well when hung upside down. Gather the fresh material in small, loose bunches with the heads at different levels to prevent crushing, and to allow air to circulate freely. Hang large flower heads individually. Tie the bunches with rubber bands, which will keep them secure as the stems dry and shrink, and hang them where they will not be disturbed.

You can also air dry some plants, such as grasses, in an upright position, using a tall, empty juice can. Weight the can with sand or stones so it

Right and **below** Air-dry flowers upside-down or upright in a warm, dry, well-ventilated location.

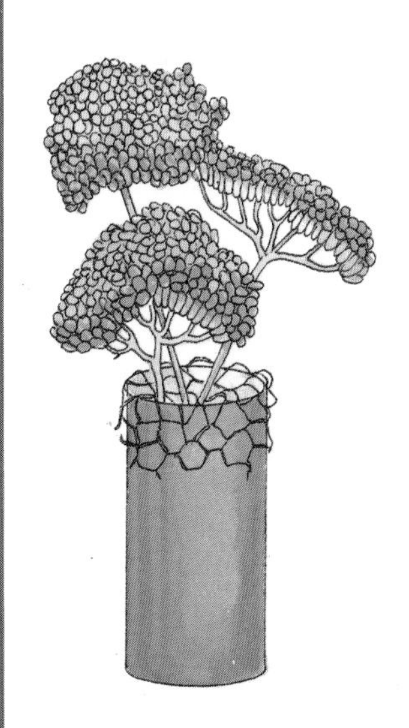

doesn't tip over. Cover the top of the can with chicken wire, available at hardware stores, and insert the stems through the chicken wire, avoiding overcrowding. If some species wilt before they dry, try fresh spêcimens and put one inch of water in the bottom of the can before inserting the stems. The water will eventually evaporate, and the plant material will continue to dry without wilting.

Store air-dried material in cardboard boxes, loosely wrapped with a thin layer of silica gel in the bottom, covered by paper. Individual flowers or small bunches should be loosely wrapped in tissue or newspaper.

Drying with desiccants

A desiccant is material that absorbs moisture from the air. Silica gel, sand, borax, and cornmeal are desiccants used to preserve plants, so they retain their natural shape and color. Specimens are placed in an airtight container, covered with desiccant, and sealed. By cutting holes in the bottom of the container and covering them with tape before filling it with desiccant, the tape can be removed at the end of the process, so the drying agent trickles out without disturbing fragile specimens. For a container without an airtight lid, plastic wrap and waterproof tape will form a suitable seal. Use newly opened flowers collected on a dry day. Cup-shaped, tubular, or many-petaled flowers are best dried face-up, others, face-down. Tall flower spikes may be dried vertically in a tall cup or horizontally in a long container. To shut the container, you may have to cut the stems, leaving two inches to mount the dried blooms with floral tape and wire, available from florists' shops. Once dry, remove desiccant adhering to flowers, using an artist's brush or a photographer's lens brush.

Silica gel is available from hardware stores, pharmacies, hobby and craft shops, and chemical supply companies. It is a crystalline substance available in several textures, but the finely ground, sandy form is best. It is expensive, so buy in bulk if you can. Buy at least 5 pounds; it can be reused indefinitely by drying it in an oven between uses. White or pink granules indicate that the silica gel contains moisture; spread them on a shallow baking sheet and bake in an oven for 20-30 minutes at 250°F (120°C). When they turn bright blue, they are completely dry and ready to reuse. Store silica gel in an airtight container when not using it. Flowers dry in 2-5 days in silica gel.

Fine, washed sand is also effective, but too heavy to be used alone except with sturdy, heavy-petaled flowers. A mixture of 1 part sand to 2 parts silica gel, borax, or cornmeal is more useful. Plants may take up to 3 weeks to dry in sand mixtures.

Borax cannot be used alone because it cakes when damp and flowers left in it too long form burn

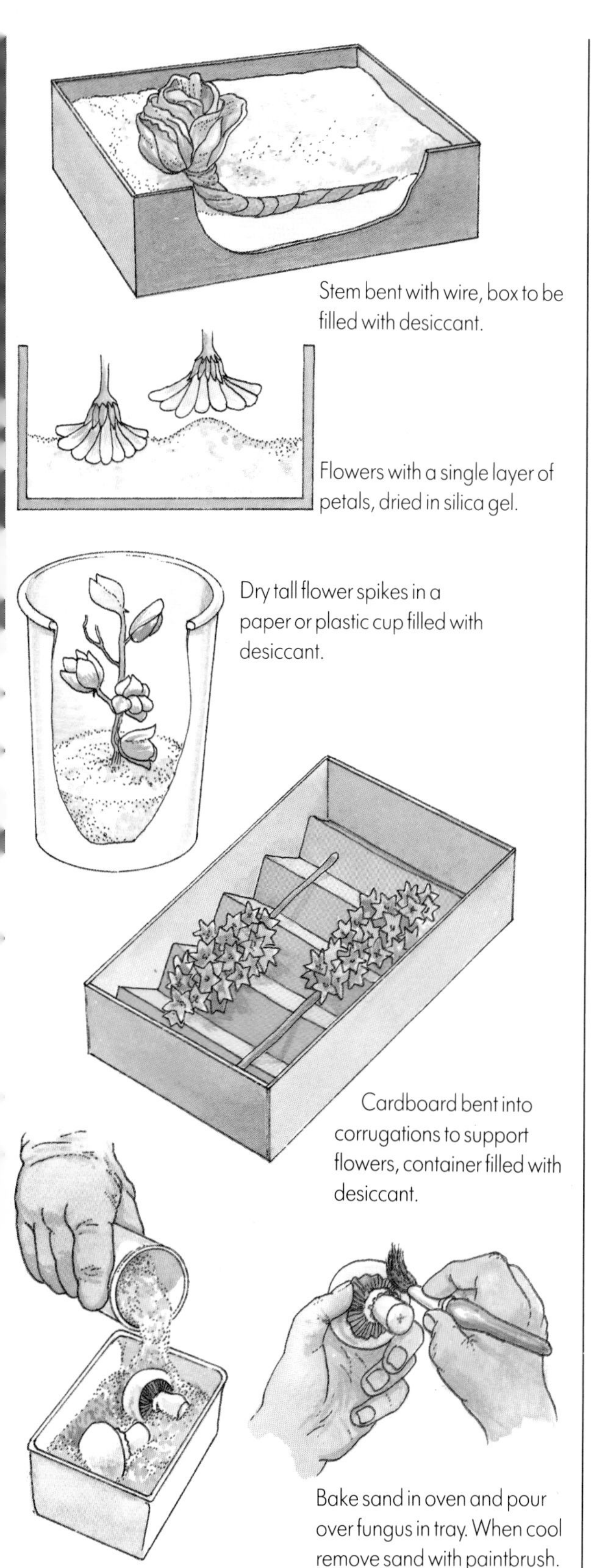

Stem bent with wire, box to be filled with desiccant.

Flowers with a single layer of petals, dried in silica gel.

Dry tall flower spikes in a paper or plastic cup filled with desiccant.

Cardboard bent into corrugations to support flowers, container filled with desiccant.

Bake sand in oven and pour over fungus in tray. When cool remove sand with paintbrush.

spots. It may be mixed with 1 part sand to 2 parts borax, or mixed in equal parts with cornmeal, a lightweight combination useful for delicate specimens. Flowers in borax mixtures will dry in 2-7 days.

Store flowers dried in desiccants in airtight containers, with a layer of silica gel on the bottom covered with paper. Use opaque containers, or place them in a dark cabinet, and seal the box with freezer tape for added protection.

Glycerine infusion

Glycerine infusion is excellent for preserving branches of mature foliage or individual leaves, but generally does not work well with flowers. Material preserved by this method lasts indefinitely and is much less fragile than living or dried plants, but its natural color is usually replaced by various brown tones. Glycerine is sold in pharmacies, but can be bought more cheaply in bulk from chemical supply companies.

Prepare a solution by mixing equal parts glycerine with boiling water, or 1 part glycerine with 2 parts water, which is cheaper but takes a bit longer to work.

Cut foliage for preserving in glycerine in late summer or early fall. Cut material at midday after several sunny days, so it is as dry as possible. Use branches up to 24 inches long. If the leaves droop, put the branch in warm water; if they do not become turgid within a couple of hours, they won't take up glycerine, and should be discarded and replaced with fresh. Strip the bark from the lower 2 inches of woody stems or crush them with a hammer. Place the branches upright in a narrow bottle, jar, or vase containing 4 inches of water/glycerine solution. If top-heavy, anchor the containers in a larger container with sand or gravel. Store in a cool, dark place, adding more glycerine solution as needed; never let the glycerine run out. When absorption is complete, in 4-28 days, leaves feel supple and oily, their veins are dark, and their color changes to shades of brown or bronze. Any remaining glycerine solution can be reused.

Store glycerine-preserved material in a dry, well ventilated spot, out of direct sunlight. It may be stored between layers of tissue paper in cardboard boxes, with air holes cut for ventilation and a few mothballs to deter insects. If beads of glycerine form on the foliage, carefully wipe them off, then wash in warm water and dry with a soft cloth.

URBAN BOTANY

PLANTS OF DISTURBED SOILS

A freshly dug garden does not stay that way for long. Small green shoots begin to appear within a few days; after another week, they evolve into recognizable plants and more shoots sprout between them. If left unchecked, by the end of the season the soil surface will be covered in greenery without a seed ever having been sown. Where did they all come from?

They are "weeds," the scourge of gardeners everywhere, and their attack on a garden is no accident. Weeds are the Boy Scouts of the plant kingdom, for their motto is "Be prepared." In the example given, some weed seeds were already in the soil, awaiting their opportunity. Others moved in afterwards; dispersed from their parent plants by a variety of means, they found conditions to their liking. Still others sprouted from the rootstocks of their deposed predecessors. They struck fast, putting down roots and sending up shoots as quickly as possible in an attempt to beat the competition.

While this scenario may sound anthropomorphic, it is not far from reality. Certain plants are particularly adapted to take advantage of disturbed soil, whether the result of human activity or some natural event. Disturbed soil is a clean slate; it provides a fresh seed bed with no established competitors for available water, sunlight, and nutrients. But competition is never far behind, and exposed soil becomes a racing ground on which plants vie to become established first and ultimately reproduce and pass on their genes to the next generation.

The practical botanist thinks of weeds, not as pests, but as plants growing where they are not wanted. While weeds seldom pose serious problems, they are often among the most readily available subjects for practical botanists to study, especially as they tend to congregate around areas of human activity.

Look for "weeds" along roadsides, field edges and railroads, and in or around fairgrounds, meadows, construction sites, lawns, embankments, playgrounds, and gardens.

Some common species of disturbed areas

common dandelion	daisy fleabane
Asiatic dayflower	bluets
red clover	bachelor's button
crown vetch	common burdock
Japanese honeysuckle	meadow goldenrod
field mustard	common mallow
black mustard	catnip
teasel	jimsonweed
common mullein	Queen Anne's lace
butter-and-eggs	tall buttercup
birdsfoot trefoil	bouncing Bet
rocket larkspur	white campion
dame's rocket	scarlet pimpernel
charlock	mouse-ear chickweed
shepherd's purse	common morning glory
lady's thumb	tiger lily
yarrow	coltsfoot
spotted knapweed	oxeye daisy
periwinkle	curly dock
Deptford pink	New England aster
common St. John's wort	common milkweed
common chickweed	sheep sorrel
rugosa rose	black-eyed Susan
day lily	common sunflower
common barberry	fireweed
orange hawkweed	butterfly weed
chicory	field bindweed
bull thistle	purslane

Above Butterfly weed is a native species adapted to disturbed areas.

Right Queen Anne's lace quickly colonizes abandoned areas.

LAWN AS HABITAT

Lawns pose particularly difficult problems to plants seeking to put down roots there. Consequently, only a few plants other than grasses are able to establish themselves in this hostile environment. A lawn's major defense against invaders is the sod layer formed by the dense fibrous root systems of grass. Many grasses spread by rhizomes, underground runners that emerge far from the parent as a tuft of grass that puts down its own roots, and sends out more rhizomes. The typical lawn is laced with miles of roots and rhizomes so tightly interwoven that they are nearly impenetrable to most other plants.

Another barrier to invaders is the complex mat of living and dead grass blades which prevent soil moisture from evaporating but also prevent most seeds from reaching the soil to germinate. Most seeds that breach this obstacle are thwarted by the sod layer below. Lawns are also mown periodically, so any plant that establishes itself must be able to survive at or below the mowing height and also reproduce vegetatively, or quickly bloom and set seeds, between mowings.

Many successful interlopers in lawns produce leaves in a basal rosette, a wheel-shaped arrangement of overlapping leaves that radiate outward from the plant's base. This configuration provides the plant with several advantages over the surrounding grass. Leaves in a basal rosette tend to arch over neighboring blades of grass, depriving them of sunlight while receiving maximum exposure themselves and staying low enough to avoid mower blades. Rosettes may also funnel rain striking them toward the center of the plant, from which the water soaks into the soil and is absorbed by the plant's own roots. Dandelions, plantains, and hawkweeds are lawn plants with basal rosettes.

Another strategy invoked by would-be invaders is the production of taproots. These tough anchors can penetrate the sod layer and obtain nutrients and moisture from deeper in the ground, thus avoiding competing with the well established network of grass roots.

Plants that spread by runners or rhizomes can play the same game as grasses, snaking their way through the maze of grass blades or roots until they find a chink and putting down their own roots. Wild strawberries and violets employ this method successfully in lawns.

Fungi successfully colonize lawns because they do not compete directly with grasses. Fungi mycelia intermingle freely with dead roots and grass blades, decomposing them and deriving nourishment from them. A fruiting body can be quickly erected to release spores before the next pass of the lawn mower. Mosses also may find a niche in a manicured yard, although not in the midst of the grass. Moss needs a moist environment in which to grow, but not necessarily a lot of soil, since they do not possess true roots. They often find suitable niches between the branching roots at the base of a tree, or on the lower tree trunk itself, if conditions permit.

Some common lawn visitors

white clover
common dandelion
common blue violet
common chickweed
orange hawkweed
scarlet pimpernel
selfheal
bluets
red clover
wild strawberry
common plantain

Many pioneering plants attempt to colonize lawns, but cannot survive periodic mowing or flower between mowings. To demonstrate this, allow a small patch of lawn to grow for a month, or summer, and observe. How many new plants are you able to identify?

Less-manicured lawns may host a surprising variety of botanical visitors. Four examples are, clockwise from upper left: common strawberry which spreads by above-ground stolons; four sky-blue petals and yellow centers identify the flowers of low-growing bluets, which spread by creeping underground rootstocks to form dense colonies; common plantain thwarts competing grasses by shading them with a dense basal rosette of broad, spoon-shaped leaves; common blue violets are a favorite find – their edible leaves are sometimes used in salads, and delicious jelly and syrup can be produced from their blossoms.

PARKS

Parks are excellent places for botanizing. Most state parks and all national and provincial parks contain the same plant species that inhabited the area before the creation of the park, and accurately portray the flora in that region. National forests and national wildlife refuges are also valuable public resources for the botanist.

City and municipal parks are often rich in plant life, but vary considerably in terms of diversity and in their use of ornamentals (selectively bred species which may or may not resemble their wild relatives). Many parks are formally landscaped, while others strive for a more natural look through the use of indigenous plants. Nearly all parks, however, prohibit the removal or destruction of plants, so plan on "collecting" only with a camera.

Left Parks frequently offer urban botanists the best opportunities to examine a large number of species. Botanical gardens, such as the New York Botanical Garden shown here, offer an added advantage, for most of their specimens are labeled, and interpretive information is presented along the trails and at the visitor's center.

Above To experience subtropical vegetation that is easily accessible, Bok Tower Gardens in Lake Wales, Florida is a fine place to visit.

Right Members of the lily family, like this day lily, are popular in municipal parks. Perennials such as these are often planted because they do not need to be replaced every year as do annuals.

Meet-a-tree

Parks are places where people can play, and one game that is particularly fun for practical botanists is called Meet-a-tree. Originally conceived to teach children to tune in to their senses of touch, smell, hearing, and taste, it works equally well for adults. All you need is a blindfold and a partner. One partner should blindfold the other and spin him or her around several times, then lead the person on a roundabout course to a tree. Seeing partners are the eyes for blindfolded ones, and must guide them safely, alerting them to hazards and directing their steps to avoid obstacles. At the chosen tree, place his or her hands on the trunk and step back to give the person room to explore, then ask the following questions:

Touch

- How does the bark feel? Does the texture change as you move your hands up the trunk, or out the limbs?
- How big is the trunk? Can you reach all the way around and grab your other hand? Is there more than one trunk? Can you reach the lowest branches, and if so, how high are they?
- How do the leaves feel? Is their texture on one side different from that on the other? What shape is the leaf? How big is it? What are the edges like?
- Are the roots exposed? How many can you feel?
- If the sun is shining, which side of the tree is warmer?
- What shape and size are any buds?

Smell

- Scratch the bark and sniff. What does it smell like? Does the bark of a twig or root smell different from that of the trunk?
- Crush a leaf and notice its smell.
- What does its fruit, if any, smell like? Crush one if necessary.

Taste

- If the tree produces edible fruit, give your partner one to sample, but be absolutely certain of your identification.

Sound

- Rap the trunk with your knuckles and note the sound.
- Crumple a leaf in your fingers and note the sound.
- If there is a breeze, try to distinguish the sound it makes as it moves through the tree. Some trees make peculiar creaks and groans as the wind moves them.

When the person feels well acquainted with the tree, lead him or her on another twisted course away from it, remove the blindfold, and have him or her try to find that tree again. Have them do a quick sketch of the tree first, and see how closely it resembles the real thing. Afterwards, reverse roles.

AGRICULTURAL LANDS
ALIEN VS. NATIVE

Plants can be placed into one or two broad categories which tell the practical botanist very much about them. "Native" plants are those which evolved in North America in the habitats in which they are largely found today. The New World has been relatively undisturbed over the millennia and these plants are accustomed to the slow pace of nature. Although our continent has long been populated by Indians, they changed little of their environment. Any alterations made were on a relatively small scale and often mimicked the forces of nature, such as using fire to open up forest and create habitats attractive to game. These minor variations were not disruptive to the native flora as a whole, so there was little pressure on plants to adapt to radical environmental changes.

Plants in the other major group are known as "alien," "introduced," or "exotic" species. These did not evolve in North America, but were imported accidentally or purposely. Most of our alien species originated in Europe or Asia where they spent thousands of years adapting to the disruptions and alterations associated with civilization. They are the survivors, the opportunists able to take advantage of the variety of new ecological niches created by human activities. Their aggressive nature makes them the scourge of gardeners and landscapers, who wage all-out war against them annually. We know them as the "weeds" of lawns, gardens, roadsides, and anywhere soil and native vegetation has been disturbed.

Native plants, by contrast, have only been exposed to civilization for two or three hundred years at the most, and the majority have only been subjected to widespread development within the past century. In the evolutionary time frame, this is an extremely short period, and native plants in general have not been able to keep up with the pressure to adapt. Consequently, many native species do not compete well in developed areas. All threatened and endangered plant species in North America are native, and habitat loss is the main threat to their existence.

Many alien plant species that have successfully established themselves in North America originated in regions of Europe and Asia where agriculture had been practiced extensively for thousands of years, and consequently they thrive in the farmlands of this continent that mimic their "native" habitat. The conditions they require to complete their life cycles include copious direct sunlight, moderate and regular precipitation, a temperate climate, and periodic soil disturbances, such as tilling, to prepare a fresh seedbed for the next generation. Interestingly, many of these (oxeye daisy, bull thistle, common dandelion, Queen Anne's lace – to mention a few) were introduced accidentally when their seeds were harvested along with crops and sent to the New World in grain shipments. Others, like some clovers, were planted here as crops themselves but escaped cultivation and now grow wild.

Above New England aster is one of the relatively few native species that thrive in such managed areas as meadows, thickets, and roadsides.

Right Grasses do particularly well in agricultural areas. Timothy is an import that escaped cultivation as a hay crop and now grows wild.

MEADOWS AND PASTURES

Meadows and pastures are both areas cleared of brush and trees for agricultural purposes. Pastures are grassy areas where livestock are grazed, while hay meadows are artificial grasslands used to produce hay. Abandoned fields and meadows naturally exhibit a greater array of wild species than actively worked ones. These habitats should not be confused with the alpine meadows of the high country, mountaintop "balds" of the Appalachians, and other areas where meadow-like conditions are maintained semi-permanently by environmental factors such as soil type and moisture, fire, elevation, and temperature. Such natural meadows tend to support a greater diversity of species and more native varieties than agricultural meadows.

Farmers plant hay meadows with crops such as timothy, rye, clover, and alfalfa that can be harvested and fed to their animals during winter. Clover and alfalfa are legumes which, as well as supplying winter forage, enrich the soil by "fixing" atmospheric nitrogen in their root nodules. These meadows are mown annually and crops are rotated regularly, to prevent certain soil nutrients being depleted from continually supporting the same crop. Hay meadows, therefore, are largely dominated by one species, but interlopers establish themselves on a regular basis.

Above Red clover is a very common plant of meadows and pastures. A hay and forage crop native to Europe and Asia, it escaped cultivation on this continent.

Pastures are subjected to different conditions, and support a differing array of plant life. To survive in a pasture, a plant must be able to withstand or avoid grazing, and recover from trampling by heavy, hoofed animals. Grasses grow from their roots, unlike most vascular plants, which grow from their tips, so grasses suffer no permanent harm when their tips are sheared off by cows, horses or sheep. This same action would kill or seriously inhibit the growth of most other plants. (Over-grazing, when grasses are cropped to the ground and denied the chance to recover, is destructive.) Defenses against grazing include the piercing spines of thistles, common pasture inhabitants; other plants have evolved stinging hairs or extremely distasteful foliage to help them thwart livestock.

Left Meadows like this one near Sudbury, Ontario, are excellent botanical sites, yielding native and alien species.

Some common plants of meadows and pastures

(Check your field guide to determine which occur in your area.)

Wildflowers

curly dock
common dandelion
tall buttercup
oxeye daisy
black-eyed Susan
selfheal
harebell
bachelor's button
common sunflower
daisy fleabane
Indian blanket
meadow goldenrod
New England aster
common milkweed
wild mint
red clover
California poppy
butter-and-eggs
purple coneflower
butterfly weed
mountain phlox
Mexican hat
scarlet pimpernel
sulphur flower
shooting star
prairie blazing star
common ragweed
chicory
lance-leaved coreopsis
bluets
bull thistle
orange hawkweed
pearly everlasting
spotted Joe-Pye weed
yarrow
wild blue flax
wild geranium
common blue-eyed grass
Canada lily
Turk's cap lily
catnip
black mustard
wild blue phlox
wild strawberry
Indian paintbrush
evening primrose
Queen Anne's lace
fivespot
Chinese houses
tidy tips
rocket larkspur
dame's rocket
purple prairie clover
common mullein

Grasses

timothy
little bluestem
meadow fescue
sheep fescue
blue grama
side oats grama
quack grass
Kentucky bluegrass
broomsedge
wild oats
switchgrass
big bluestem
buffalo grass
orchard grass
bent grass

Below Trees, such as eastern red cedar, will invade pastures if given the chance. Clumps of grass grow on manured soil.

FENCEROWS, THICKETS, AND WOODLOTS

Fencerows, which often define individual fields and meadows or delineate the boundaries of neighboring farms, are an entirely separate habitat from the open spaces just a few feet away, and contain the shrubs and trees excluded from most fields. Depending on the orientation and width of the fencerow, there may be shade-tolerant wildflowers, normally associated with deciduous woodlands. At the core of many fencerows, especially in the Northeast, may be a stone wall that hosts a variety of lichens. The spaces between the stones also tend to create cool, moist microhabitats that favor the growth of mosses and ferns.

Thickets

Thickets are patches of uncultivated land, less linear than fencerows and ranging from a few dozen square yards to acres. They generally have been cleared in their recent history, as shown by their young trees and characteristically large amount of brush. Once cleared, they may have been found too moist or too rocky for cultivation, or they may have been overlooked for other reasons. They are excellent examples of the shrub-sapling stage of succession in your area. Shrubs such as raspberry, blackberry, common barberry, multiflora rose, hawthorn, sweetbrier, and flowering dogwood may abound. Trees present are usually mature pioneer species, such as gray birch and eastern red cedar of the eastern United States, or saplings of the next seral stage. Some dense thickets contain multi-stemmed saplings of bigtooth or quaking aspen, indicating that their rootstocks remained alive under the thicket while it was in use, held in check only by grazing, mowing, or other events.

Woodlots

Woodlots are old thickets or forest remnants surrounded by agricultural land or, more recently, by housing developments. Nearly all low-lying land in the east has been cleared at some time since the New World was colonized, but those that have been left alone for, say 75-150 years, have undergone secondary succession and reached a stage approximating the mature climax forest characteristic of the region. The distinction between a forest and a woodlot is arbitrary, but woodlots are generally considered small, wooded areas of less than a few acres and located at least several hundred yards from other wooded areas. Larger woodlots, especially those close to other large wooded tracts, tend to develop an interior region that biologically resembles a true forest, which is covered in the chapter on woodlands.

Woodlots usually exhibit the same vegetation zones as a forest: canopy, understory, shrub layer, herb layer, and ground layer. The canopy is the uppermost vegetation and is composed of mature trees of the locally dominant species. The understory consists of smaller, shade-tolerant trees and sun-loving trees rushing to fill any holes in the canopy left by fallen trees. The next lower layer consists of shrubs, many of which were present in earlier seral stages. The shrub layer is particularly evident near the southern edge of woodlots, where more direct sunlight is likely to penetrate. Below this is the herb layer, consisting of woodland wildflowers from the woodlot's interior, and field and meadow species toward the woodlot's edges. Finally, the ground layer is home to low-growing mosses, liverworts, club mosses, and fungi.

Left Fields in agricultural areas are frequently bordered by fencerows, or windbreaks. These areas are havens for species that cannot withstand regular plowing or mowing.

Below Multiflora rose (center) and common barberry (bottom) are common plants of eastern agricultural areas, each capable of forming impenetrable thickets. In many places they are considered weeds.

Some common plants of fencerows, thickets, and woodlots

staghorn sumac
black raspberry
multiflora rose
hawthorn
gray birch
black mulberry
black walnut
black locust
chokecherry
eastern white pine
tulip tree
evening primrose
white baneberry
bluets
virgin's bower
common burdock
Jerusalem artichoke
Philadelphia daisy
closed gentian
cut-leaved toothwort
large-flowered trillium
Turk's cap lily
common milkweed
scarlet sage
field bindweed
nodding ladies' tresses
fringed loosestrife
turtlehead
spotted touch-me-not
Japanese honeysuckle
poison ivy
hedge bindweed
haircap moss
red raspberry
common barberry
sweetbrier
flowering dogwood
eastern red cedar
fox grape
shagbark hickory
sweet gum
pin cherry
sassafras
northern red oak
red baneberry
mayapple
Canada anemone
wood anemone
goldenrod
New England aster
spotted Joe-Pye weed
wild geranium
Canada lily
smooth Solomon's seal
butterfly weed
Oswego tea
wild bergamot
bittersweet nightshade
yellow-fringed orchid
spring beauty
Asiatic dayflower
Virginia creeper
trumpet honeysuckle
downy serviceberry
bracken fern

Where to go

Agricultural lands with fencerows, thickets, and woodlots tend to occur in the eastern half of the continent, and especially in the Northeast. Nearly all are privately owned, but most farmers are friendly and accommodating if you introduce yourself, explain your interest, and ask permission to look around. Remember to leave everything as you found it: close gates that were closed, leave open gates open, don't disturb livestock, and walk *around* planted crops, not through them. The farmer might appreciate hearing about what you've found and receiving a personal thank-you and a photograph.

QUADRATS AND TRANSECTS

Quadrats and transects are useful techniques for learning about composition of and relationships within plant communities. Identifying plants with field guides and keys is important but it is only the first step in studying natural history. The next step is to understand the significance of an organism's presence in a particular place. Recognizing different plant communities and analyzing them with quadrats and transects helps you predict roughly which plant species you are likely to encounter there, as well as other characteristics, including soil composition, moisture content, drainage, wildlife species and climate.

Quadrats

A quadrat is a sample plot of a given size, usually square, but occasionally rectangular or circular. Quadrats may be single units or divided into subunits, with size depending on the characteristics of the vegetation under study. Quadrats of one-fifth of an acre are often used to study forest canopy trees, while understory trees and shrubs may be evaluated with smaller plots. Quadrats of one square yard are adequate for evaluating herbaceous plants, such as grasses and wildflowers, especially in a meadow or other habitat with dense vegetation.

The accuracy of any study depends on the number and size of quadrats used, simply because the more quadrats taken, the more it increases the chances of encountering members of the community. In a relatively uniform habitat, the increase in accuracy diminishes after about five or six samples, assuming that they are randomly chosen, so it is within the first few samples that most of the information is accumulated.

Types of quadrats

List Quadrat The plants found are listed by name. A series of list quadrats provides an overview of the species present. Dividing the number of quadrats in which a particular species was found by the total number of quadrats investigated and multiplying by 100 shows the percentage of that species' occurrence within quadrats. (Example: goldenrod occurred in 25% of all quadrats in Meadow X.)

Count Quadrat The names and numbers of each species are recorded, showing the basic composition of the community and the abundance

The optimum size of a quadrat varies with the subjects studied: a quadrat analyzing data on large subjects, such as trees (**left**) should be larger than one analyzing meadow plants (**bottom left**). Quadrats and transects (**below**) yield different types of data; you may need to use both methods to draw accurate conclusions.

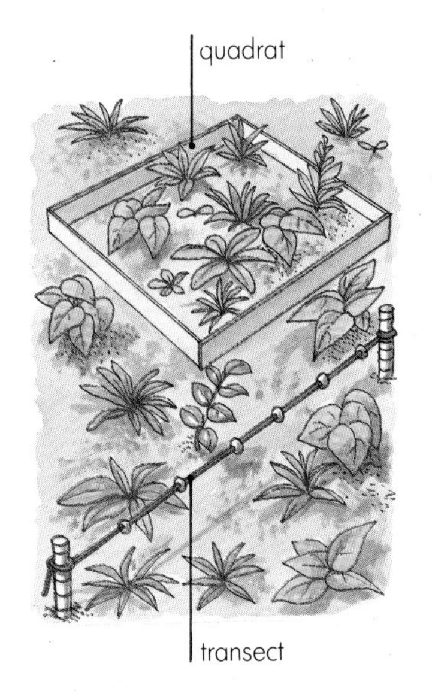

of individuals. (Example: The species composition of Meadow Y is 25% timothy, 15% little bluestem, 20% goldenrod, 15% milkweed, 10% Queen Anne's lace, 6% New England aster, 5% yarrow, and 4% tall buttercup.)

Cover Quadrat The percentage of ground surface shaded or covered by vegetation is recorded, either by species or for total vegetation. Estimate the area shaded or covered by each plant, total the estimates for each species, divide by the area of the quadrat, and multiply by 100 to obtain the percentage of coverage of each species.

Chart Quadrat Draw a scale map of the quadrat showing the locations of individual plants. This is a tedious job and is useful only to the researcher doing a long-term study of vegetation changes.

Transects

Line Intercept Observations are made on a line or lines laid out randomly or systematically across the area under scrutiny. As with quadrats, accuracy increases with the number of lines surveyed.

- Stretch a tape measure between two points 50 or 100 yards (or meters) apart.
- Divide the line into intervals, such as 1 yard, 3 yards, or 5 yards.
- Move along the line, recording for each interval the species touched by the line or lying directly over or under it, as well as the distance their foliage covers along the line.

- Summarize the data by:

a the number of intervals in which each species occurs along the line.
b the frequency of occurrence for each species relative to the total number of intervals sampled.
c the total distance covered by each species along the transect.
d the total distance covered by vegetation along the line and the total uncovered distance.
e the total number of individuals of each species along the transect.

From this data, the following calculations can be made, their results expressed as a percentage:

- Relative Density: total individuals of species X divided by total individuals of all species, times 100.
- Dominance (cover): total intercept length of species X divided by the total transect length, times 100.
- Relative Dominance: total intercept length of species X divided by the total intercept length of all species, times 100.
- Frequency: number of intervals in which species X occurs divided by the total number of transect intervals, times 100.
- Relative Frequency: frequency value of species X divided by the frequency value of all species, times 100.

The line intercept method is easy, objective, and particularly useful for measuring changes in vegetation throughout the study area. It is less accurate for estimating frequency and abundance than the quadrat method, because the probability of an individual being sampled by line intercept is proportional to its size.

Belt Transect An effective combination of the quadrat and the line intercept method in which a "belt", or long, narrow unit, is laid out across the area under study and divided into segments of equal size. These segments may then be used as quadrats and surveyed. This is an accurate method for determining abundance, frequency, and distribution, especially when more than one belt is utilized. It also shows vegetation changes along a gradient, such as from dry to moist soil.

Plant profiles

Making a plant profile (a two-dimensional representation of the plant community of a given area) is an excellent way to illustrate the changes in a plant community along an environmental gradient, such as increasing or decreasing moisture, sunlight or temperature. Lay out line intercepts in an area with a definite environmental difference from one end to another, such as a meadow with distinctly dry and wet areas. Analyze your findings according to the methods listed above, but do the calculations for only half of the transect at a time to determine whether the plants you've found exhibit a preference for one extreme over another.

If you notice a definite change in the plant profile while doing line intercepts, challenge yourself to determine the environmental factor causing it.

ABANDONED FIELDS AND SUCCESSION

An abandoned field becomes a laboratory in which to study a classic example of secondary succession. In the spring following the last harvest, the field begins to green up on its own. Some plants are germinating from seed dropped during the last harvest, and some are grasses and dicots that were growing in the field along with the crop.

Many others, however, are invaders that began their assault during the previous several years, sending in expendable troops in order to have them in place should an opportunity for exploitation arise. The so-called weeds, with their predisposition to disturbed soils, are the vanguard of the invasion. Milkweeds, dandelions and thistles drifted from great distances. Others, such as common burdock and flax, hitch-hiked in on the fur or feathers of passing mammals and birds. In some cases, the seeds in fruits eaten by mammals and birds passed through their digestive tract before being deposited in the field. Periodically, seeds and fruits stockpiled by mice and other rodents are forgotten or abandoned, and they, too, germinate. Then there are those that spread slowly but surely by means of underground rhizomes.

During the first couple of years, opportunistic annuals and biennials tend to predominate. The exact species composition of pioneer communities is highly variable, depending on seed sources in the area, farming practices prior to abandonment, moisture content and soil condition. With each passing year, conditions improve as more organic matter is added to the soil which, along with increased shading, reduces moisture evaporation and soil temperatures, allowing plants with stricter requirements to establish themselves. Native perennials are better adapted to utilize the improved conditions, and with their shading and spirited growth, quickly crowd out the annuals. The short life cycles of annuals were advantageous at first, but once native perennials move in, annuals cannot compete with the perennials' earlier spring growth and longer life cycles.

An old field evolving into a mature climax community of the type common to that area is a marvelous opportunity for a practical botanist to witness the development of plant communities firsthand. Return several times each year throughout the growing season and make detailed notes about the appearance and disappearance of species as the field ages. Return when the plants go to seed and examine them to deduce their method of dispersal and how they found their way there. The knowledge amassed with a long-term study of secondary succession is both fascinating and invaluable.

Left The first plants of any stage of ecological succession to invade an area are called pioneer species. Eastern red cedar is typically among the first trees to invade abandoned fields in eastern North America.

Above Ecological succession will occur on nearly any new or disturbed area, but the classic example is an abandoned field like the one shown here. The so-called "weeds", well-adapted to open, disturbed soils, move in first, followed by perennial shrubs, saplings, and finally, trees.

Left Sumacs, such as this staghorn sumac, are also pioneers of the shrub-sapling stage. Staghorn sumac is easily identified in the fall and winter by its fuzzy, dark red seeds growing in dense, erect cones.

GRASSLANDS

GRASSLAND AS A COMMUNITY

Until quite recently, many people interested in nature saw the prairies only as a great void to be crossed to reach the mountains and coasts on either side of the continent. Its landscape, a vast sea of grass and farmland, seemed so flat and boring that its biological wealth was ignored. These grasslands, in fact, are one of the largest and most productive ecosystems in North America. They are so productive that their conversion to cropland has threatened their existence in some regions, while their vastness and seemingly monotonous nature led them to be taken for granted.

Despite their name, grasslands support much more than just grasses. They host an abundance of forbs (herbs other than grass), desert plants, shrubs, and occasional trees, the latter particularly abundant along rivers and streams. Additionally, they provide habitat for a surprising array of mammals, birds, insects, and reptiles. Grazing and burrowing animals, such as bison, pronghorns, and prairie dogs, are especially well suited for prairie life. Intact grasslands are a stable, climax ecosystem that seems to defy the principles of ecological succession, but other forces of nature are at work here.

Types of grasslands

"Grasslands" here covers the tallgrass prairie, mixed prairie, shortgrass prairie, and intermountain grasslands. One might also argue for the inclusion of alpine meadows, tidal wetlands, and desert grassland, but, while consisting largely of grass-like plants, these habitats are so radically different in composition and origin that they are covered in other chapters.

Grassland characteristics

When grasslands are mentioned, one normally thinks of the prairie, which is indeed the largest grassland area in North America, beginning with the tallgrass prairie east of the Appalachians and blending gradually into the shortgrass prairie at the foothills of the Rockies. There are other types of grasslands in the East, Southwest, and West which are examined in other chapters.

Most grasslands share certain characteristics. With the exception of some eastern, intermountain, and California grasslands, most occur on flat or gently rolling terrain. Prairie soils, while originating from different bedrock sources, are all slightly alkaline, rich in humus, deep, and very fertile. Precipitation averages between ten and thirty-nine inches per year, but is punctuated by periods of severe drought. The climate in general is marked by high rates of evaporation.

Grassland communities are stratified into three zones: roots, ground layer, and herbaceous layer. The root layer is the most pronounced of any major ecological community, with some grass species reaching six feet or more into the soil. At least half of the plant is represented underground, and virtually all of it in winter. When they finally die after supporting the plant for 20 to 50 years, the

Left Pawnee National Grassland, Colorado, is one of many areas set aside to preserve these botanically and ecologically fascinating habitats.

Right Yellow goatsbeard, a member of the sunflower family distinguished by long, pointed bracts extending beyond the yellow ray flowers, is a common grassland resident.

decomposing roots leave an intricate system of miniscule tunnels that supply air and water to the deeper layers of topsoil, a benefit to the surviving plants in the immediate vicinity. The ground layer in taller grasslands is characterized by lower temperatures, lower light levels, and higher humidity than occur in the upper herb layer exposed to drying winds and solar radiation. At ground level a layer of mulch forms, made of dead stems and leaves that decompose as they come into contact with the mineral soil. This mulch adds humus to the soil, increasing its water-retention capability. The herbaceous layer develops as the season progresses, with ground-hugging plants growing and blooming first, eventually hidden as taller plants form a middle layer and, finally, an upper layer of very tall grasses and wildflowers that bloom in the fall.

A word about grasses

Grasses are unique among flowering plants. The grass family, or *Poaceae*, is the third largest family of flowering plants. They are truly ubiquitous, with representatives on every continent.

Grasses are monocots, and as a group they are easy to recognize, but because of their inconspicuous flowers, they have a reputation of being difficult to identify as individual species. Their ecological and economical value is indisputable, though. Much of our food comes

Left Tallgrass prairie, the most fertile of the midwestern grasslands, has succumbed to the plow throughout most of its range. Today it exists only on carefully preserved or restored tracts and on overlooked areas, such as along railroad rights-of-way.

shortgrass prairie
mixed grass prairie
tallgrass prairie
intermountain grasslands

directly or indirectly from grasses and most of our meat-producing and dairy livestock are fed grain or graze on grass which we cannot digest.

Physiologically, grasses differ from most other flowering plants in that their region of active growth is at the leaf base, near where it joins the stem, not at the tip. This enables them to survive, and even thrive, with moderate grazing, which allows sunlight to penetrate farther, sometimes making for a healthier plant. In another interesting adaptation, new blades of grass are hidden inside the hollow stems of older shoots, able to take over if the older shoot is clipped off.

To survive extreme heat, drying winds, drought, fire, heavy grazing, and long winters, grasses have developed massive root systems, which in some species are estimated to collectively equal up to three miles in length per plant. Grasses spread by underground rhizomes as well as by seed, and often these rhizomes, along with their fibrous root system, form dense tangles that lock the topsoil into a solid layer called sod, making it difficult for many would-be competitors to put down their own roots. Living blades of grass weave a tangled mat with the dead leaves of previous seasons that further retards invading plant species. Those grasses that are not sod-formers are called bunch grasses and, true to their name, grow in bunches with other plants occupying the spaces in between.

NATURE'S FORCES AT WORK

The existence of grasslands as a stable ecosystem seems to be a direct contradiction to the normal course of ecological succession. Whether primary or secondary, one group of plants normally colonizes a given area and, in doing so, gradually alters the environment, creating conditions less favorable for their own progeny but more hospitable to the pioneer species of the next seral stage. The grass stage normally gives way to a shrub-sapling stage of sun-loving woody perennials, which is in turn succeeded by a young forest. This will ultimately be replaced by a mature climax forest.

There are many forces at work maintaining the grasslands. Because of the deep and extensive root systems which compose a large percentage of the entire plant, grasses do better in semi-arid conditions than do many trees and shrubs. Although grasslands receive moderate amounts of precipitation annually, their flat terrain poses no obstacle to drying winds, so much of their moisture gain is lost to evaporation. This is more pronounced in the warmer south, where more precipitation is needed to maintain the same type of grassland vegetation as in the north. Precipitation in grasslands is often concentrated in peak seasons interspersed with periods of drought. Finally, much more precipitation falls on the windward side of mountain ranges than on the leeward side, and thus a "rain shadow" effect occurs on the eastern slopes of the Sierra Nevada, Cascade, and Rocky Mountain ranges. The grasslands that occur here are composed of short, sparse grasses. As precipitation increases and evaporation decreases to the east, taller and lusher grasslands develop until finally the eastern forests appear.

Fire is another natural force keeping grasslands in check. Regular lightning-caused fires fanned by constant winds have always been part of grassland ecosystems, clearing organic debris that shades the soil, returning nutrients to the soil, and creating a fresh seedbed for wildflowers. Grass roots remain unscathed by the rapidly moving fire and, with the energy stored in their extensive root systems, resprout almost as soon as the ashes are cool. However, most trees and shrubs native to these regions are less tolerant of fire, and they suffer serious setbacks as a result.

One major force at work on the prairies used to be the great herds of bison and other herbivores. They roamed the prairies by the millions and their grazing stimulated renewed grass growth after they left, leaving tons of manure to fertilize the soil and sustain the vigorous new growth. Also, chunks of sod torn up by their thundering hooves left bare soil, in which annual and perennial wildflowers could establish themselves. With their population reduced to a mere shadow of its former grandeur, bison and other herbivores are no longer a significant factor in the grassland ecosystems.

Below The great North American grasslands lie in rain shadows, areas on the leeward side of major mountain ranges. As moist air masses move from west to east, they are forced upward as they strike the mountain ranges. As the moist air moves higher, it cools below its dew point, clouds form, and precipitation falls on the western slopes. Cresting the mountains, the air descends, warming quickly. Due to the moisture lost, the air is much drier, precipitation is scarcer, and a rain shadow is created.

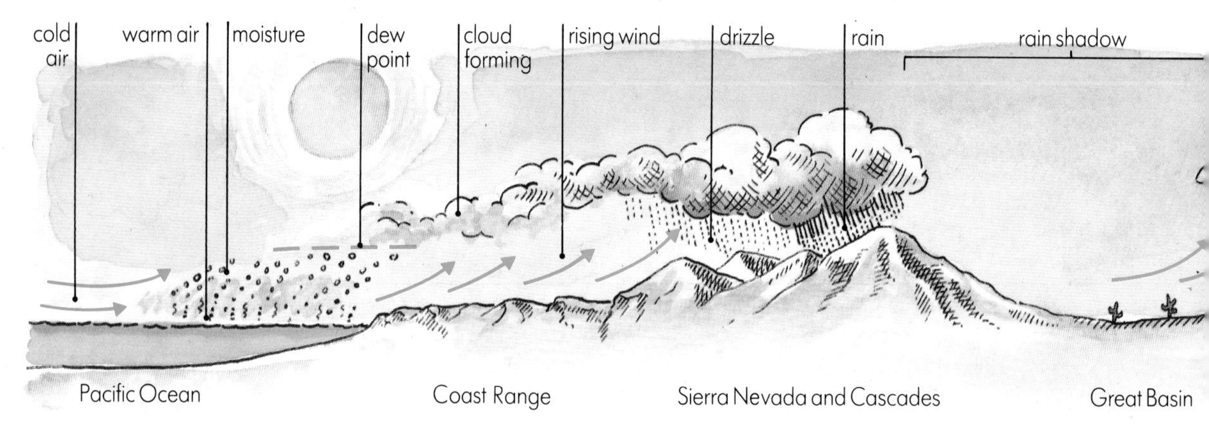

Above Fires such as this one in the Everglades, are equally important in maintaining the prairie grasslands, especially in the tallgrass prairie where it keeps trees at bay.

Left These bison are a remnant of the great herds which used to exercise a regulating force on the prairies.

TALLGRASS PRAIRIE

There was a time in the not-so-distant past that the tallgrass prairie covered more than 400,000 square miles in 12 states. It was the first extensive grassland encountered by pioneers as they moved west, and it left them truly in awe. They had seen huge meadows and pastures before, but nothing like this. There are accounts of grasses so high that they could conceal a man on horseback. Where one could see over them, the monotony of the scene was overwhelming; there was seemingly nothing but a sea of grass that stretched to the horizon, an analogy that may have prompted their Conestoga wagons to be dubbed "prairie schooners."

Sadly, nothing remains of this spectacle but a few scattered nature preserves and overlooked pockets of native prairie vegetation. Ironically, pioneers first thought prairie soils were infertile because trees were absent. In fact, the area received substantial precipitation, and it was the periodic fires set by lightning or Indians that kept the forests at bay. They soon realized their misconception and set about cultivating massive areas of fertile soil, aided by a product of the Industrial Revolution, the Prairie Queen. This tempered steel plow, produced by John Deere, was sturdy enough to rip through the dense black sod with the aid of strong beasts of burden.

Tallgrass prairie is not always tall. Winter snows and wind beat the lofty vegetation nearly to ground level, and the new grass shoots do not reach their maximum height until mid-July. In the meantime, the low-growing forbs of the area bloom first in the spring, followed by progressively taller wildflowers, and finally the taller grasses produce their plumes of less-conspicuous flowers in the late summer.

Tall grasses, including big bluestem, switchgrass, Indian grass, and, in wetter areas, slough grass dominate this system. Despite the preponderance of grasses, some trees also grow along riverbanks and in upland areas. Cottonwoods, green ash, box elder, and American elm follow the watercourses, while bur oak is common in drier upland areas, developing a four-foot taproot in its first year in order to compete with the deep-rooted grasses.

Below Tallgrass prairie, displaced throughout most of its range by agricultural fields, receives enough precipitation to support a forest. Regular fires maintained it as a grassland for hundreds of thousands of years.

Prairie blazing star (**above**) is a common wildflower of the tallgrass prairie, which is defined by a predominance of big bluestem grass, (**left**).

Some common plants of the tallgrass prairie

Grasses

big bluestem
Indian grass
prairie cordgrass
needlegrass
little bluestem
bluejoint
redtop
switchgrass
slough grass
foxtail barley
grama grass
prairie dropseed
timothy
Kentucky bluegrass

Wildflowers

compass plant
goldenrod
bluets
two-flowered Cynthia
alumroot
prairie false indigo
yarrow
boneset
prairie blazing star
white sweet clover
ragweed
horse nettle
sundrops
shooting star
long-headed thimbleweed
star-flowered Solomon's seal
marsh pea
white snakeroot
pale lobelia
rattlesnake master
goat's rue
black-eyed Susan
swamp saxifrage
prairie coreopsis
rough blazing star
pineapple weed
smooth aster
common milkweed
prairie phlox
downy painted cup
Canada anemone
blue flag
smaller redroot
northern bedstraw
hoary puccoon
wild geranium
spreading dogbane
butterfly weed
Indian blanket
Indian paintbrush
New England aster
purple prairie clover
spotted Joe-Pye weed
showy evening primrose
Maximilian's sunflower
New York ironweed
spiderwort
marsh marigold
bird-foot violet
common sunflower
death camas
Jerusalem artichoke
pasqueflower
smooth aster
evening primrose
plains larkspur
giant sunflower
crazyweed
prickly pear

Trees and shrubs

pasture rose
black oak
eastern cottonwood
quaking aspen
leadplant
American plum
prairie crabapple
bigtooth aspen
green ash
Siberian elm
shrubby cinquefoil
common chokecherry
bur oak
common barberry
eastern red cedar
paper birch
smooth sumac
post oak
hackberry
common prickly-ash

MIXED GRASS PRAIRIE

As one proceeds west, the mixed grass prairie marks the beginning of the Great Plains. It also divides the tallgrass prairie, where trees grow if not precluded by fire, from the shortgrass prairie, where they will not. Most trees found in the mixed grass prairie are riparian species (willows, cottonwoods, elms, hackberry, and green ash) common along watercourses. The few trees that grow elsewhere in the mixed grass prairie are found in the eastern part of this grassland.

Unlike tallgrass prairies to the east which are maintained primarily by fire, the grassland habitat in mixed grass prairies is largely determined by its low annual precipitation, ranging between fourteen and twenty-three inches per year. Wind speed and the rate of evaporation are also higher than on tallgrass prairies, which increases the region's aridity.

Mixed grass prairie is dominated by little bluestem, needlegrass, Junegrass, and, in the northern reaches, western wheatgrass. Also common are buffalo grass and blue grama, the two dominant species of the shortgrass prairie, which, in this region, occur more frequently at higher (and therefore drier) elevations. Big bluestem is also present, mostly along river terraces and other moist areas. Due to the shorter height of the vegetation, sunlight is better able to reach lower plants in mixed grass prairie than in tallgrass prairie. Thus, mixed grass prairie exhibits a more layered plant community in mid-summer when the grasses are fully grown than does the tallgrass prairie.

Left As the name implies, mixed grass prairies exhibit a mixture of both shortgrasses and midgrasses. Though this grassland is usually at least partially green through the summer, severe droughts result in parched scenes like this in Badlands National Park, South Dakota.

In combination with others, little bluestem (**below**) is one of the indicator species of the mixed grass prairie. Purple prairie clover (**bottom**) is just one of the common wildflowers of this region.

Left Leadplant, so named for the dense gray hairs covering its foliage, is one of the most common shrubs of the mixed grass prairie.

Some common plants of the mixed grass prairie

Grasses

little bluestem
Junegrass
blue grama
switchgrass
big bluestem
Kentucky bluegrass
Indian grass
redtop
needlegrass
western wheatgrass
buffalo grass
prairie cordgrass
foxtail barley
needle-and-thread
timothy

Wildflowers

Indian paintbrush
tall goldenrod
birdfoot violet
sego lily
spreading dogbane
prairie rose
plains larkspur
evening primrose
bladder campion
death camas
silverleaf scurf pea
giant sunflower
compass plant
blue salvia
rough blazing star
prairie blazing star
many-spined opuntia
spotted Joe-Pye weed
Jerusalem artichoke
pale agoseris
crazyweed
hairy golden aster
pasqueflower
butterfly weed
wild blue flax
black-eyed Susan
horse nettle
mouse-ear chickweed
Carolina anemone
prairie false indigo
common sunflower
snow-on-the-mountain
Indian blanket
locoweed
purple prairie clover
prairie smoke
prickly pear
stiff goldenrod
prairie gentian
common milkweed
prairie mimosa
Maximilian's sunflower
Mexican hat

Trees and shrubs

leadplant
shrubby cinquefoil
eastern cottonwood
bur oak
osage orange
hackberry
Russian olive
common prickly-ash
prairie acacia
smooth sumac
eastern red cedar
American plum
green ash
Siberian elm
common chokecherry

SHORTGRASS PRAIRIE

Proceeding west from the mixed grass prairie one comes to the shortgrass prairie, an area of light, infrequent precipitation, low humidity, and high winds. These factors combine to produce an extremely high rate of evaporation, second only to that of a true desert. Due to these arid conditions, a permanent dry zone exists deep in the subsoil, into which plant roots do not penetrate.

The two dominant grasses on the shortgrass prairie, buffalo grass and blue grama, have adapted well to such a demanding environment. Although their roots are shorter than most other prairie grasses, they account for more than 90% of each plant! Both are sod formers by nature, but can also grow in isolated clumps in drier areas. Buffalo grass is easily recognized by the above-ground stolons, or runners, it sends out to put down roots and colonize new areas, and blue grama is characterized by its arched, one-sided flower stalks. They concentrate their growth in the cooler, moister weather of spring and fall, becoming dormant during the heat of midsummer, once they use all available moisture in the soil. (This characteristic distinguishes shortgrass prairie from mixed grass and tallgrass prairies, where moisture is generally retained in the soil year-round.) When this happens, the grasses lose their brief greenery and take on a parched, yellowish-brown appearance, emphasizing the dryness of the area.

Due to their severe climate and reduced humus, shortgrass prairies have been largely unaltered by agriculture. They are, however, susceptible to overgrazing by livestock. Unlike their wild counterparts, the bison and pronghorn, which are constantly on the move and give grasses and forbs a chance to recover between grazings, livestock is liable to be confined to one area for long periods. Their close cropping weakens plants' root systems, so that other grasses better adapted to constant grazing, such as Kentucky bluegrass and cheatgrass brome, are able to invade, changing the character of the shortgrass prairie. Severe overgrazing, along with the removal of cattle for slaughter, may permanently destroy the prairie by breaking the nutrient cycle of growth and decomposition. And without the protection of the moist, shady mulch created by decomposing grass, the soil bakes in the sun, becoming susceptible to the wind erosion that occurred in the Dust Bowl of the 1930s.

As one travels west across the prairies and into the rainshadow of the Rockies, plants begin to exhibit more and more desert adaptations protecting stems and leaves, the plant's main water storage organs from desiccation. Among the more common are light-colored leaves that reflect heat and light; dense, hairy or wooly stems and leaves that insulate them from temperature extremes and reduce the flow of moisture-robbing winds over the surface, and a waxy or resinous coating on the foliage that reduces transpiration.

Right Many people associate cacti only with deserts, and are quite surprised to find them in grasslands as well. In reality, grasslands often receive only slightly more precipitation than do deserts. Plains pricklypear is one of several species common to the prairies.

Some common plants of the shortgrass prairie

Grasses

buffalo grass	blue grama
needle-and-thread	Kentucky bluegrass
little bluestem	western wheatgrass
Junegrass	red three-awn
foxtail barley	

Wildflowers

yellow bell	white prairie clover
purple prairie clover	blackfoot daisy
cowpen daisy	Tahoka daisy
western wallflower	mule's ear
common sunflower	Indian blanket
arrowleaf balsam root	curlycup gumweed
Hooker's evening primrose	pasqueflower
sego lily	silverleaf scurf pea
wild blue flax	plains wallflower
Rocky Mountain bee plant	prairie gentian
western pink vervain	many-spined opuntia
locoweed	wooly locoweed
crazyweed	death camas
vase flower	plains larkspur
scarlet globemallow	buffalo gourd
plains yucca	desert plume
hairy golden aster	field milkvetch
camphorweed	dotted blazing star
plains pricklypear	candelabra cactus
Mexican hat	

Trees and shrubs

fringed sage	snakeweed
Rocky Mountain juniper	prairie sage
Siberian elm	rabbit brush
four-wing saltbush	eastern cottonwood
common chokecherry	osage orange
quaking aspen	Russian olive
winter fat	

Left The shortgrass prairie extends eastward from the foothills of the Rocky Mountains and grades into the mixed grass prairie. Grassland boundaries are neither distinct nor permanent, but shift back and forth with changes in climate and precipitation patterns.

INTERMOUNTAIN GRASSLANDS

Flanked by the Cascade and Sierra Nevada ranges to the west and the Rockies to the east, the intermountain grasslands cover the majority of Utah and Nevada, the eastern third of Oregon and Washington, and a small portion of southern Idaho. They encompass the Great Basin, a huge area with no drainage to the sea, in which most streams and rivers drain into the Great Salt Lake.

Although considered a grassland, one of the dominant plants is not a grass, but sagebrush. The grasses found here are bunchgrasses that space themselves widely by putting out horizontal as well as vertical roots to obtain moisture. A shortage of summer rainfall favors cool-season species, such as bluebunch wheatgrass and Idaho fescue, which are dormant during summer. Unfortunately, in many areas bluebunch wheatgrass has been displaced by cheatgrass brome, an exotic invader that capitalized on the cultivation of wheat in the northern part of the intermountain region and spread from there. The overgrazing that accompanied the cattle boom of the 1880s weakened the bluebunch wheatgrass, hastening its replacement with cheatgrass brome, which had evolved

Below One of the dominant forms of vegetation in the intermountain grasslands is big sagebrush. Grasses and herbs reign in the spaces between the sagebrush plants.

Left The intermountain grasslands are not always vast expanses of grass, as shown in this view near Long Creek, Oregon. Depending upon the topography, they may occupy only narrow bands on mountain slopes between the dry, desert areas at lower elevations and the pinyon/juniper woodland or lodgepole pine forest at higher, moister elevations.

Left Bluebunch wheat grass is an indicator species of the intermountain grasslands. Unfortunately, in some areas this native has been largely displaced by cheatgrass brome, an alien species accidentally imported from Asia with wheat shipments.

Some common plants of the intermountain grasslands

Grasses

Idaho fescue	bluebunch wheatgrass
Kentucky bluegrass	cheatgrass brome
foxtail barley	bottlebrush squirreltail

Wildflowers

scarlet globemallow	fragile pricklypear
hood phlox	longleaf phlox
prairie star	arrowleaf balsam root
threadleaf groundsel	cream cup
desert paintbrush	scarlet gilia
locoweed	hoary cress
yellow bee plant	field milkvetch
vase flower	spreading dogbane
mule's ears	pale agoseris
sego lily	many-spined opuntia
yellow bell	broom snakeweed
elegant camas	prairie smoke
pale wallflower	western wallflower
hawk's beard	sweet fennel
fiddleneck	desert plume
flatpod	crazyweed

Trees and shrubs

big sagebrush	winter fat
desert sage	greasewood
shadscale	rabbitbrush
snakeweed	little horsebrush
four-wing saltbush	Mormon tea
Utah juniper	Russian olive
singleleaf pinyon	curlleaf cerocarpus
Siberian elm	threadleaf phacelia
iodine bush	

to withstand the year-round grazing of camels and wild horses on the steppes of Asia, whence the contaminated shipments of wheat originated. Overgrazing also led to an increase in sagebrush, which in many places grows more densely now than it did in past centuries.

The vegetation of this region varies with changes in elevation and latitude. Only ten to fifteen inches of precipitation falls annually, with higher elevations receiving more than the basins. Consequently, plants less tolerant of drought are found higher on the mountain slopes. South-facing slopes, which receive the most direct sunlight, are warmer and therefore drier than northern slopes, so the orientation of the mountainsides also affects the types of vegetation. In the northern intermountain zone, sagebrush grassland occupies the basins and lower mountain slopes, but to the south it remains only on the lower mountain slopes, giving way to true desert plants in the southern basins, the hottest and driest parts of this region. Forbs occur in the intermountain grasslands, but not in the abundance in which they are found throughout the prairies.

WOODLANDS

THE FOREST AS A COMMUNITY

At first glance, forests, with their myriad species, may seem indecipherably complex. Complex they are, but all forests follow orderly patterns that govern their structure and function. Just as an ornithologist must be able to recognize birds of prey before learning to distinguish broad-winged hawks from goshawks, the practical botanist should concentrate on recognizing general ecosystems before trying to unravel the intricacies of any particular one.

Forests, like all ecosystems, are primarily composed of a given set of species that are consistently present wherever that particular habitat is encountered. Of course, there are always exceptions, variations, and overlap between various habitats, but generally speaking, the combined presence of a few key *indicator species* tells you which type of habitat you are observing. By first narrowing possibilities, you reduce your task to a manageable level. By knowing what other species you might expect to encounter, you increase the likelihood of actually finding them. For instance, if you are in a forest dominated by American beech and sugar maple, you may correctly surmise that you are in the northern hardwood forest, the northern tier of the eastern deciduous woodlands, and can expect such companions as yellow birch, striped maple, hobblebush and eastern hemlock. More importantly, it eliminates a host of other plants that you should not expect to find here, such as sweetgum or magnolia.

An ecosystem functions like one giant organism, and like any organism, it has a definite structure and orderly functions. Forests have a structure that is particularly easy to discern. Typical forests are stratified vertically, having five distinct vegetation zones, although one or more of these may be reduced or absent in certain forest types. From the top down they are the canopy, understory, shrub layer, herb layer, and ground layer. The canopy, of course, consists of the tallest trees, while shade-tolerant trees make up the understory and shade-tolerant shrubs, the shrub layer. Ferns and ephemeral spring wildflowers are the primary constituents of the herb layer, and the ground layer, which may be considered part of the herb layer, hosts low-growing plants like mosses, liverworts, fungi, and club mosses.

In a temperate deciduous forest, winter is the dormant period when most trees shed their leaves and enter a resting state. In spring, the plants do not come into leaf all at once. Instead,in a very orderly

Decomposers, like this fungi (**left**) known as chicken-of-the-woods, play a crucial role recycling nutrients in a forest. Interactions among different species such as this are what makes a forest an orderly community and not just a disjointed assemblage of organisms. Showy lady's slippers (**right**) are important as well as beautiful members of the forest community; deducing their significance is part of the fun of botany.

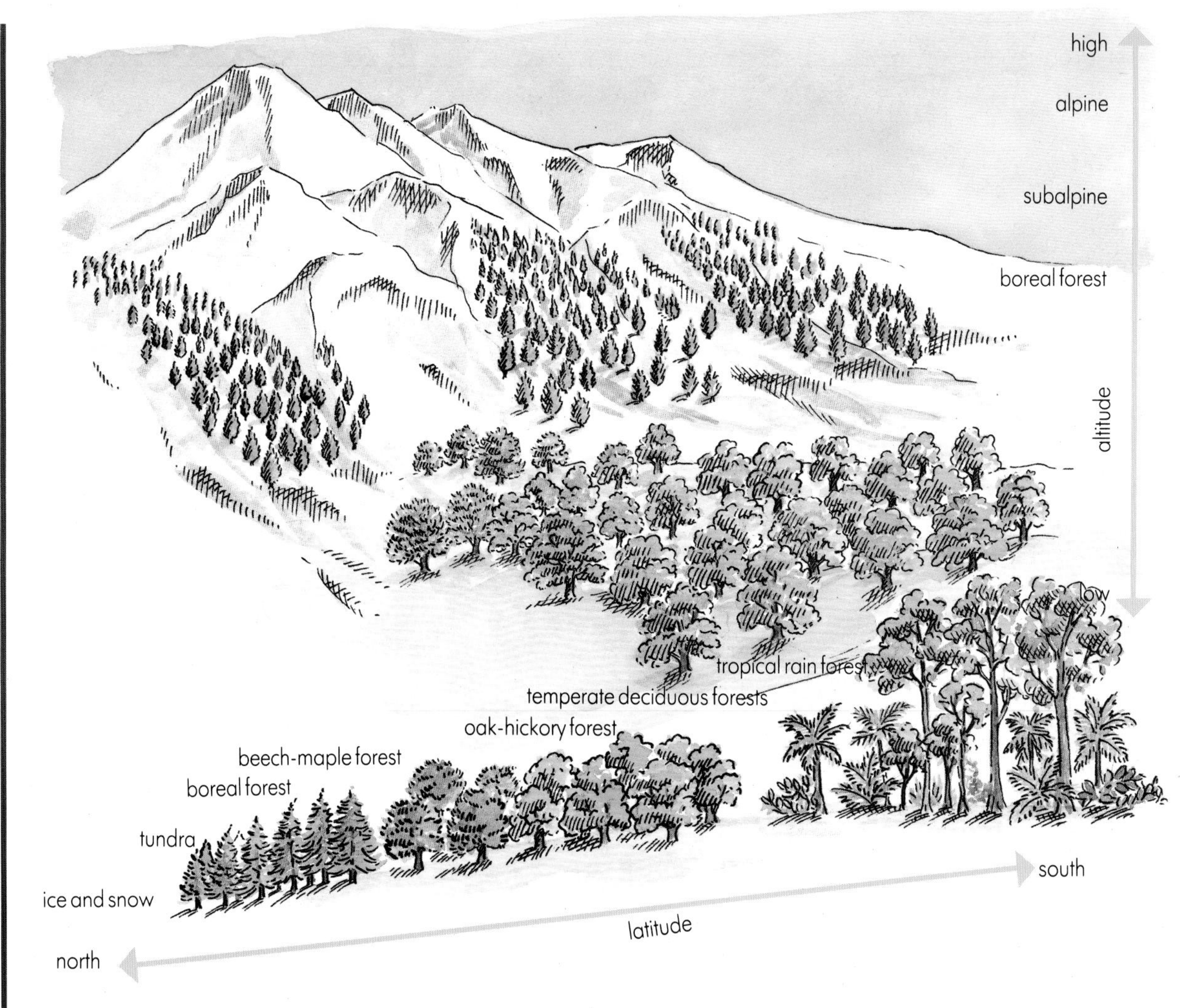

Above Plant communities change with elevation as well as with latitude. Climbing a few thousand feet in elevation from the base of a North Carolina mountain will take you through the same plant communities as would walking 1,000 miles or so due north.

procession, the herb layer develops and flowers first, followed by the shrub layer, the understory, and finally the canopy. This is no accident. As a matter of survival, herbaceous plants of deciduous forests evolved early season growth as their means of securing enough solar energy to reproduce and develop food reserves to carry them through another year. Shrubs also need an early start for the same reason, but the selective pressure on them is not as great as that affecting herb-layer plants, so shrubs can wait until a bit later in spring to produce leaves. Thus, the zone with the greatest biological productivity is the one that came into leaf most recently. Once the canopy's leaves are fully formed, it becomes the forest's primary zone of photosynthesis for the rest of the season. It has been estimated that once the canopy of a typical eastern deciduous forest is in full leaf, as little as 1% of the available direct sunlight reaches the ground, the other 99% being absorbed or reflected by the canopy leaves.

There are exceptions to this model of forest stratification. Both xeric (very dry) and hydric (very wet) deciduous forests tend to have more open canopies, so stratification is less evident and the orderly pattern of coming into leaf from bottom to top is not so critical. Dark, dense coniferous forests dominated by evergreen needleleaf trees often have reduced or absent zones underneath the canopy because there is little opportunity for those plants to garner the sunlight they need.

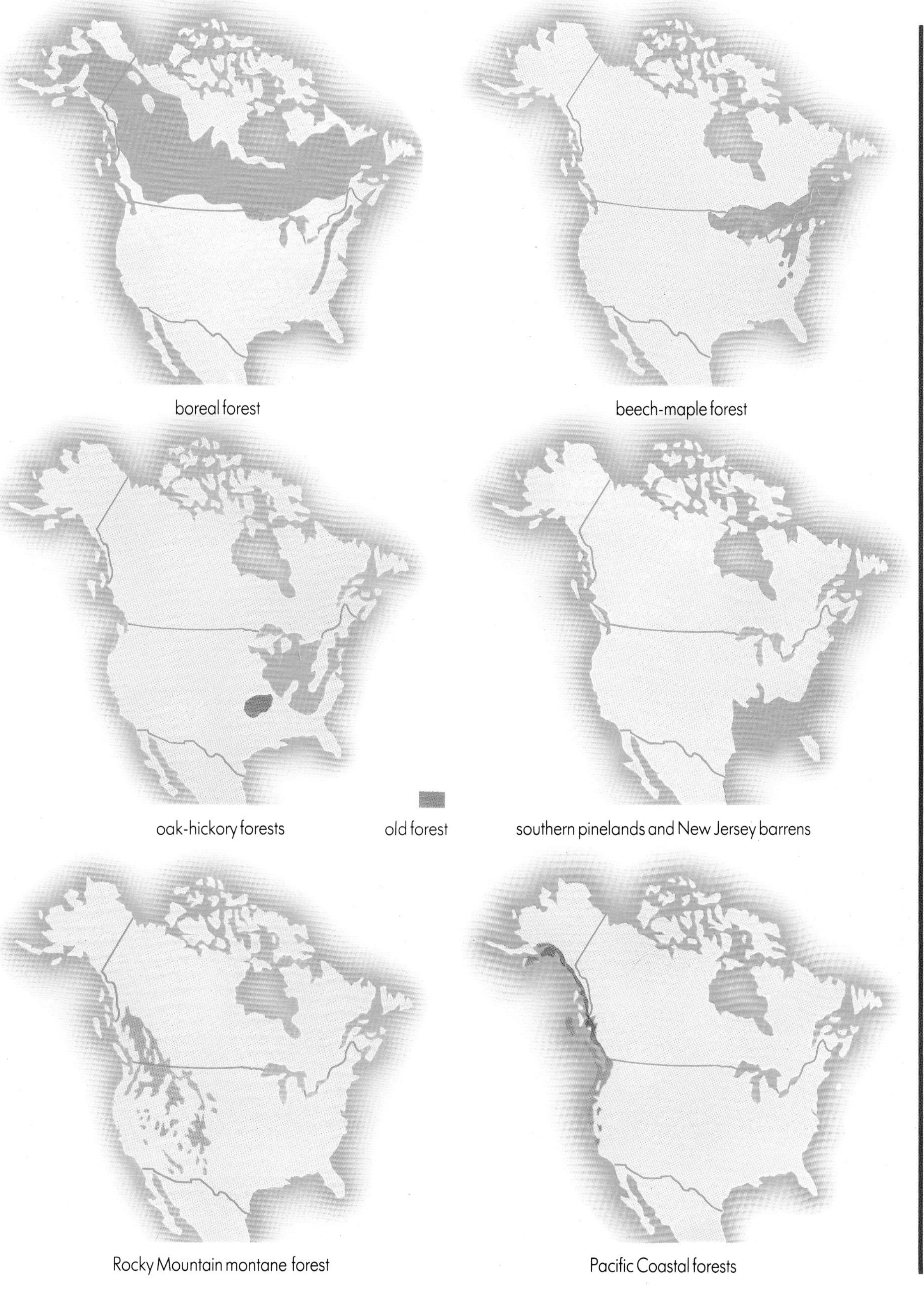

boreal forest

beech-maple forest

oak-hickory forests

southern pinelands and New Jersey barrens

Rocky Mountain montane forest

Pacific Coastal forests

FIRE AND FOREST ECOSYSTEMS

Many people have grown up with the image of Smokey the Bear depicting the catastrophic effects of forest fires and warning that only *we* could prevent them. Countless scenes on television and in the movies also portrayed fire as a major destructive force in nature that robbed us of our natural resources.

As it turns out, Smokey the Bear and his message may have been the forests' own worst enemy. Fire, in fact, is an integral part of many temperate forest and grassland ecosystems. In a forest, plants grow, die, and decompose as they are recycled to build the soil and to provide nutrients to future generations of plants. Often, there is a lag period between the time a tree dies and when it actually hits the ground, where active decomposition takes place. It may stand for years as a dead snag, providing valuable nesting sites and food sources for wildlife. When it finally does fall, it may get caught on other trees, so it is still suspended in the air. Branches usually fall or get knocked off separately, and their irregular shape often keeps them off the ground. As long as dead wood is kept from contact with the moist soil, circulating air continues to dry it.

Since the accumulation of this debris occurs faster than the decomposition of the material under it, a mechanism is needed to periodically clean out the forest floor without cheating the soil of the nutrients due it. Periodically, perhaps every 10 to 25 years, droughts desiccate the debris even further, fires are sparked by lightning, and a "cool" ground fire sweeps through the forest, devouring the downed "fuel" and depositing a nutrient-rich ash on the forest floor, which provides a fresh seed bed for new growth. Fanned by a breeze, this type of fire usually moves fast, barely lingering long enough to scorch the bark of mature trees, which many times suffer no harm at all. Moving slowly against the wind, the fire may indeed kill some trees, creating a natural meadow and providing the edge habitat critical to so many wildlife species before

In many cases, forest fires actually improve the health of the forest, clearing debris and returning nutrients to the soil. Even areas in Yellowstone National Park that were devastated by the wildfires of 1988 (**top**) have begun their recovery (**bottom**).

Above Many plants thrive in the wake of fire. Here, quaking aspens sprout prolifically from their rootstocks after a fire, and lodgepole pines produce many seed cones sealed with resin that require the heat of a fire to release their seeds.

succession guides it back to a mature forest. Eventually the fire reaches a firebreak, such as a stream, road, or a recently burned area, and is extinguished for lack of fuel. All is well.

Once it was decided that fires were evil, everyone set about snuffing them out with a vengeance. Fire towers kept watch across our nation's forests, and at the first wisp of smoke an army of firefighters would descend upon the flames and extinguish them before they could properly be called a forest fire. Meanwhile, the forest continued producing new wood, trees continued to die at their normal rate, and the fuel piled up. Firefighters found it harder and harder to contain the blazes. The accumulating flammable material, combined with periodic droughts, created a tinderbox that, when ignited, led to a crown fire that raced through the canopy, killing trees, leaping firebreaks and outpacing the efforts of firefighters. Such uncontrolled fires are commonly called wildfires and, under the most extreme circumstances, a fire storm results where desiccated trees burst into flame from the intense heat before the blaze even reaches them. These truly destructive fires are the direct result of human interference. This expensive lesson was graphically demonstrated during the incineration of Yellowstone National Park in the summer of 1988, during which nearly 1,000,000 acres in and around the park were blackened by an inferno fueled by 80+ years' worth of debris and winds that topped 70 miles per hour.

Fire is now accepted as a natural part of most natural ecosystems in North America, and as essential to the health and vigor of many habitats. Areas like the Everglades and California's sequoia forests are actually preserved by regular fires. Some plants, such as the jack pine, whose cones only open in intense heat, require fire to reproduce. Hopefully, the overwhelming scientific evidence will convince us to see fire as a part of nature's process of renewal. Fire management should be left to the proper authorities; arson is both dangerous and criminal.

BOREAL FORESTS

Up north, where winters are long and people are scarce, stretches the largest tract of forest in North America, the boreal forest. It covers much of Canada and Alaska, from the U.S. border to the arctic treeline and from the Atlantic coast of Newfoundland to the Bering Sea. It is primarily a coniferous forest, composed largely of needle-leafed, cone-bearing trees, and can roughly be divided into three east-west belts of vegetation. The southernmost pine/hemlock band is overshadowed by the middle spruce/fir zone, followed by the northernmost taiga zone, in which the firs and spruces become increasingly dwarfed until they finally give way to arctic tundra.

Because of the harsh climate, the diversity of life in the boreal forest is rather low compared to other North American ecosystems; only the arctic tundra supports fewer species. What it lacks in diversity, however, the boreal forest makes up in sheer numbers. At first glance, the forest seems totally dominated by black spruce, white spruce, and balsam fir, but closer scrutiny reveals other trees. Red, white, and jack pines are common in the southernmost regions. Jack pines thrive in poorer, dry soil, colonizing such areas first and eventually yielding to red pine. Quaking aspen, balsam poplar, bigtooth aspen, and paper birch, the primary deciduous species, inhabit clearings opened by fire or other disturbances.

In much of the boreal forest, interlaced evergreen branches block the sunlight so effectively that the understory, shrub, and herb layers are greatly reduced or missing entirely. Other barriers to undergrowth are the thick, spongy carpet of discarded needles, and the nutrient-poor soil made acidic by decomposing needles. Sphagnum moss often carpets the ground in wetter areas, and lichens and mushrooms abound. Wildflowers include starflower, bluebead lily, bunchberry, bearberry, twinflower, and northern white violet. Other wildflowers and shrubs are largely restricted to clearings and the edges of lakes and streams where sunlight is available. Among the more common shrubs are blueberry, elderberry, sheep laurel, and thimbleberry.

Bogs and meadows, the remnants of filled-in bogs, dot the landscape. These are products of glacial activity that scoured out depressions in the ground and then deposited huge chunks of ice as they retreated, which melted and filled the hollows. Though surrounded by the boreal forest, these bogs are unique communities, covered in the Wetlands chapter (see pp113-15).

An extension of the boreal forest reaches south to Georgia along the highest peaks and ridges of the Appalachian Mountains. Here, white spruce and balsam fir are replaced by red spruce and Fraser fir, but otherwise it bears many of the hallmarks of a boreal forest.

The boreal forest (**above**) is the largest forest community in North America. Composed largely of coniferous trees, its dense canopy admits little light, thus the dark, moist forest floor supports a relatively low diversity of plant species. Two of the most common trees of the boreal forest are white spruce (**far left**) and balsam fir (**left**).

Some common plants of the boreal forest

Trees and shrubs

black spruce
balsam fir
red pine
bigtooth aspen
balsam poplar
red maple
sheep laurel
thimbleberry
mountain maple
white pine
white spruce
jack pine
quaking aspen
paper birch
tamarack
mountain ash
blueberry
elderberry
green alder

Ferns and mosses

sphagnum moss
oak fern
bracken fern

Wildflowers

bluebead lily
bunchberry
twinflower
nodding trillium
northern white violet
harebell
yellow lad's slipper
common wood sorrel
purple-fringed orchid
calypso
orange hawkweed
goldthread
Canada mayflower
red baneberry
one-sided pyrola
starflower
bloodroot
pink lady's slipper
shinleaf
pinesap
partridgeberry
fireweed
bearberry

Two species that thrive in the dimly lit herb layer of the boreal forest are starflower (**above**) and bunchberry, a member of the dogweed family (**left**).

NORTHERN HARDWOOD FOREST

Also known as the transition forest because it shares many species with the cool, dark, moist boreal forest, and continues to the warmer, drier oak-hickory forests to the south, the northern hardwood forest is the northern tier of the eastern deciduous woodlands. It also lies between the boreal forests of the higher Appalachian Mountain peaks and the oak-hickory forests at their bases. The winters here are shorter and milder and the growing season longer than in the boreal forest, and since its canopy is more open than the boreal forest, the northern hardwood forest also supports richer understory, shrub, and herb layers.

Much of the canopy is dominated by American beech or sugar maple. Interestingly, researchers have found that beech seedlings grow better under maples than they do under other beeches, and vice versa. Consequently, American beeches may usurp the canopy while the sugar maples inhabit the understory and bide their time until the current generation of beeches dies off. They then grow to fill the canopy while a new crop of beech seedlings grows underneath them, and so on. The third most common hardwood species in this region is yellow birch. In the moist lowlands, northern red oak, American basswood, and American elm are prominent.

Two needle-leaved species, eastern hemlock and eastern white pine, grow abundantly here. Eastern hemlocks are particularly plentiful in cool, moist, shady ravines and north-facing mountainsides, while white pine thrives in open or disturbed areas, together with gray birch and pin cherry. Striped maple, downy serviceberry, and

Left The predominance of American beech and sugar maple indicate that this New Hampshire scene is part of the northern hardwood forest.

Above Paper birch, with its white bark, and sugar maples are common members of the northern hardwood forest.

nannyberry are common understory species, and the shrub layer often sports hobblebush. Compared to the rather barren floor of the boreal forest, the herb layer is a botanist's delight. Many members of the lily family grace the floor of the northern hardwood forest, including painted trillium, purple trillium, large-flowered trillium, trout lily, Canada mayflower, false Solomon's seal, bluebead lily, and others. Red baneberry, white baneberry, pink lady's slipper, bunchberry, starflower, fringed polygala, and Jack-in-the-pulpit are also present.

Species composition of the northern hardwood forest, largely dependent upon local factors such as soil type, moisture, and temperature, varies from one community to the other. For instance, in the midwestern regions eastern hemlock and yellow birch diminish while white, red and bur oaks become prominent, and American basswood replaces American beech as a dominant canopy species. The best way to familiarize yourself with local variations is to write or visit state and national parks and forests and ask for their interpretive brochures and booklets on the local flora.

Some common plants of the northern hardwood forest

Trees and shrubs

American beech
yellow birch
white pine
gray birch
paper birch
nannyberry
striped maple
American basswood
tamarack
balsam poplar
quaking aspen
highbush blueberry
black cherry
pin cherry
sugar maple
eastern hemlock
northern red oak
American mountain ash
mountain maple
hobblebush
mountain laurel
downy serviceberry
witch hazel
red spruce
great laurel
red maple

Wildflowers

painted trillium
pink lady's slipper
common wood sorrel
trout lily
partridgeberry
white baneberry
starflower
foamflower
Indian pipe
false Solomon's seal
sweet white violet
wild columbine
wild sarsparilla
Jack-in-the-pulpit
downy yellow violet
purple trillium
goldthread
bluebead lily
large-flowered trillium
red baneberry
spring beauty
Canada mayflower
harebell
purple-fringed orchid
common blue violet
wild bergamot
fringed polygala
bloodroot
sessile bellwort
bunchberry

Ferns and fern allies

ground pine
bracken fern
shining club moss
interrupted fern
ground cedar
stiff club moss
maidenhair fern

Left Frequently encountered in the herb layer of the northern hardwood forest are (clockwise from top): large-flowered trilliums, trout lilies, and bluebead lilies, all members of the lily family. Each spring, members of the herb layer race to grow and flower before the forest canopy leafs out and restricts their light.

OAK-HICKORY FOREST

This is by far the largest of the eastern deciduous forests. It is sometimes known as mixed deciduous forest, for while oaks and hickories are the most plentiful trees, up to three dozen species may vie for a place in the canopy. The term refers to two forests that were formerly considered separate: the true oak-hickory forest of Arkansas, southern Missouri, and eastern Oklahoma, and the former oak-chestnut forest of southern New England, the Middle Atlantic states, and the southern Appalachian mountains. However, with the mass demise of the American chestnut after the chestnut blight fungus was accidentally imported to this country in 1906, hickories and more oaks assumed their places in the canopy. Interestingly, the American chestnut roots are blight resistant and continue to sprout and grow as an understory species before being struck down by the infection. They seem to grow larger with each effort before succumbing, which may indicate that they are developing a resistance to the blight. Perhaps some day they shall regain their rightful place as monarchs of the eastern forests.

An abundance of both oaks *and* hickories defines this forest. White oak, scarlet oak, black oak, chestnut oak, northern and southern red oaks, mockernut hickory, pignut hickory, and bitternut hickory are all good indicator species. Wetter sites often host sweetgum, shagbark hickory, red maple, and sourwood. Pioneering trees in disturbed areas include gray birch, eastern red cedar, quaking aspen, bigtooth aspen, black locust, white pine and pitch pine in the north, and slash pine in the south. The understory of the oak-hickory forest, much more highly defined than in the northern hardwood forest or the boreal forest, includes American chestnut, flowering dogwood, downy serviceberry, sassafras, hackberry, hophornbeam, witch hazel and redbud.

The shrub layer in this forest can be quite dense. Entanglements of great laurel and mountain laurel, which at times seem almost impenetrable, paint the forest pink with their spring flowers. The elegant pinxter flower and flame azalea are also noteworthy. Highbush blueberry, lowbush

blueberry, and deerberry are plentiful. The herb layer shares many species with the northern hardwood forest: violet wood sorrel, Dutchman's breeches, cut-leaved toothwort, teaberry, mayapple, common cinquefoil, sweet white violet, birdfoot violet, Jack-in-the-pulpit, spring beauty, round-lobed hepatica, rue anemone, bloodroot, yellow lady's slipper, large-flowered trillium, trout lily, wild bergamot, wood betony, and wild columbine are just a few. Ferns are richly represented, with hay-scented fern, marginal woodfern, interrupted fern, Christmas fern, and lady fern being particularly common in the uplands. Royal fern and cinnamon fern are two species often encountered in the moist lowlands. Clubmosses, including ground cedar, shining clubmoss, and tree clubmoss, are common.

Above Flowering dogwood is a major understory species in these woods. Large white bracts surround clusters of flowers before its leaves emerge in spring.

Above Among the many wildflowers common to the herb layer of the oak-hickory forest is wild columbine, which is frequently found on sunny, well-drained mountain slopes.

Left The oak-hickory is the largest North American deciduous forest, although most of it was once oak-chestnut forest prior to the onslaught of chestnut blight.

Some common plants of the oak- hickory forest

Trees, shrubs and vines

white oak	scarlet oak
black oak	northern red oak
southern red oak	chestnut oak
shumard oak	bear oak
bur oak	northern pin oak
shingle oak	blackjack oak
post oak	pignut hickory
mockernut hickory	bitternut hickory
shagbark hickory	sweetgum
red maple	sourwood
black cherry	black walnut
white ash	green ash
blackgum	gray birch
eastern white pine	pitch pine
black locust	quaking aspen
slash pine	shortleaf pine
Virginia pine	American chestnut
flowering dogwood	downy serviceberry
sassafras	hackberry
hophornbeam	witchhazel
redbud	great laurel
mountain laurel	pinxter flower
flame azalea	highbush blueberry
lowbush blueberry	deerberry
spicebush	mapleleaf viburnum
poison ivy	Virginia creeper
fox grape	

Wildflowers

Dutchman's breeches	mayapple
cut-leaved toothwort	bloodroot
violet wood sorrel	Jack-in-the-pulpit
round-lobed hepatica	rue anemone
large-flowered trillium	purple trillium
trout lily	wild sarsparilla
wild columbine	wood betony
yellow lady's slipper	teaberry
birdfoot violet	bunchberry
common blue violet	fringed polygala
harebell	pinesap
Indian pipe	starflower
sweet white violet	twinflower
spring beauty	yellow-fringed orchid
pink lady's slipper	smooth Solomon's seal
false Solomon's seal	

Ferns and fern allies

hay-scented fern	Christmas fern
marginal woodfern	interrupted fern
lady fern	royal fern
cinnamon fern	haircap moss
ground cedar	shining clubmoss
tree clubmoss	pincushion moss

SOUTHERN PINELANDS

Southern pinelands is a catch-all phrase describing the broad crescent of coastal plains in the southeastern United States, stretching from the pine barrens of New Jersey to eastern Texas. While the unifying feature of this area is a preponderance of pines, many other tree species occur, and while pines remain dominant, the species composition can vary radically from one region to another.

One reason for this is that pinelands are not a climax community, the endpoint of ecological succession, but a "fire climax," an arrested state of succession held in check by periodic forest fires. Without fire, these areas would eventually succeed to a forest dominated by oaks, beech, hickories, and magnolia. Pinelands are constantly trying to evolve, but are regularly being returned to a state favorable to the growth of pines. Historically, fires in the southern pinelands occurred naturally from lightning strikes, but more recently, pulp and paper companies, which own huge tracts of these timberlands, have adopted the policy of regular burning, to reduce excess fuel which could lead to a wildfire and threaten their investments. In turn, fire clears the ground and prepares a fresh seedbed necessary for the germination of pine seeds. Fire also prevents the broadleafed understory taking over as the pines age, and shading the sun-loving pine seedlings.

Differences in precipitation, soil character, and drainage further alter the specific composition of different plant communities. Obviously, some plants are better adapted to wet (hydric) soil, some to dry (xeric) soil, and some prefer a happy medium (mesic soil). Pineland soils are typically nutrient-poor. Since pine needles are resistant to decay, the species richness in old pine stands is fairly low because most available nutrients are tied up in standing trees and in the carpet of pine needles on the forest floor. By contrast, the species diversity in a recently burned area is much higher because the nutrients previously locked up in forest debris have been returned to the soil and made available to other plants.

The pinelands, then, are not one type of forest, but dozens. They are defined more geographically than botanically, but some species do present themselves repeatedly. Longleaf pine is the most widespread and the leader in economic importance. It is often mixed with slash pine and saw palmetto. In the northern part of this range, loblolly pine and shortleaf pine commonly occur together. Turkey oak, another fire-adapted species, often forms solid stands on the higher elevations of the coastal plains. Live oak, the quintessential symbol of the Deep South, with its massive trunk

Spanish moss (**left**) draped profusely from branches is a common sight in the southern pinelands. An indicator species of this community is the longleaf pine (**above**).

Above Another indicator species of southern pinelands is the saw palmetto, a member of the rather limited shrub layer.

and wide, spreading crown draped in Spanish moss, is intolerant of fire but thrives in wetter soils and along the coast. Post oak, myrtle oak, laurel oak, southern red oak, southern catalpa, common persimmon, and Carolina holly are also indicator trees of this area. Common shrubs include saw palmetto, shining sumac, spicebush, southern bayberry, one-flowered hawthorn, and pinxter flower. Grasses such as wiregrass, little bluestem, and yellow stargrass dominate the herb layer. Spanish moss, an epiphyte or "airplant," cascades prolifically from tree branches throughout the southern pinelands. It is not parasitic, but photosynthesizes its own food, obtains minerals leached naturally from the foliage of the host tree, and collects rainwater by means of scales covering its stems. Another epiphyte, greenfly orchid, is also endemic.

Some common plants of the southern pinelands

Trees, shrubs and vines

longleaf pine	slash pine
loblolly pine	shortleaf pine
Virginia pine	live oak
turkey oak	post oak
myrtle oak	laurel oak
southern red oak	shumard oak
black oak	blackjack oak
water oak	willow oak
overcup oak	southern magnolia
mockernut hickory	pignut hickory
water hickory	bitternut hickory
American beech	blackgum
sweetgum	sweetbay
southern catalpa	common persimmon
Carolina holly	southern bayberry
saw palmetto	shining sumac
spicebush	one-flowered hawthorn
pinxter flower	Virginia creeper
poison ivy	American sycamore
yaupon	

Wildflowers

Spanish moss	pinesap
Indian pipe	downy false foxglove
mayapple	round-lobed hepatica
bird-foot violet	common blue violet
bloodroot	Jack-in-the-pulpit
partridgeberry	pink lady's slipper
wood betony	scarlet sage
passionflower	mistletoe
crested dwarf iris	grassleaf golden aster

Grasses

wiregrass	little bluestem
yellow stargrass	

FLOODPLAIN FORESTS

Floodplain forests of low river valleys may be considered subtypes of the larger forests that surround them, but in many respects they are different enough to merit separate consideration. Often, the species composition changes drastically in the uppermost reaches of floodwaters.

Flooding is a sort of safety valve that protects communities downstream. When receiving excess drainage from a storm or the spring melt, a river confined to a deep, narrow channel would have no choice but to increase its velocity and, therefore, its power and destructive potential. The forceful waters would tear at the sides of the channel, removing everything from soil to rocks; severe currents can move boulders. The torrent would turn into a brown soup as its load of sand, silt, and clay increased. Eventually, the river would widen, the current slow and lose power, and its cargo of sediment would settle. Were this to happen at the mouth of the river where it meets the ocean, the estuary located there would be devastated. Estuaries, the nurseries of our oceans, are biologically the most productive areas on earth. Subjected to unchecked floodwaters, they would be smothered under countless tons of sediment.

Fortunately, a mechanism exists that controls the destructive potential of rivers and has indirectly resulted in the evolution of floodplain forests. As the waters rise and their velocity increases, they crest the bank of their normal channel, spreading out across a flat plain and promptly slowing down and depositing their load along the length of the river, instead of just at the end. With each flood, alluvium, a rich layer of silt full of nutrients and organic matter, is deposited on the already fertile soil of the floodplain.

The plants that grow in floodplains generally require moister and/or more fertile soil than those of the surrounding region. Of course, floodplain forest plants have had to adapt to take advantage of this bonanza; they must be able to survive periodic flooding or repopulate in the face of it.

Many trees will drown in soil that is saturated for any length of time. Green ash and willow species, all common residents along stream and river banks, are stimulated to form new, air-filled roots to replace those that die due to flooding. Lenticels, tiny openings in their bark, expand to allow more air into the trees. Under extreme conditions, these species can switch to anaerobic respiration, which does not require oxygen. Many riparian (river) species depend on the very waters that may kill the parents to distribute the seeds of the next generation. The seeds of some willows and cottonwoods, for example, actually need to be submerged in order to germinate.

Above Black willows are residents of riverbanks, lakeshores, and low, wet places in the east. They often grow right by the water's edge.

Farther back from the river bank, cottonwoods colonize levees, low ridges or mounds parallel to the river that are composed of coarse sediments deposited by the higher floodwaters. The larger size of the sediment particles means more space between them and better drainage, so that even in close proximity to the water this soil may be quite dry, and cottonwoods must put down deep taproots to protect themselves from drought.

Beyond the levees, the species composition, which varies geographically, grades into those progressively less tolerant of prolonged flooding. In the Northeast, a community of sycamore, black ash, boxelder, American and slippery elms, and red, silver, or black maples occurs after the cottonwoods, with American beech and tulip poplar just above the normal high water mark. In the Southeast, levees are often followed by overcup oak, water hickory, waterlocust, and red maple, which in turn may be replaced by sycamore, sweetgum, water oak, willow oak, and Nuttall oak.

In this region, the highest floodwaters may see shagbark hickory, post oak, and swamp chestnut oak. In the arid West, floodplain forests may represent a distinctly different community from the surrounding area. Willows and cottonwoods still abound along the river banks. Floodplains in the Rockies accommodate quaking aspen, balsam poplar, and blue spruce, while in the Southwest they host tamarisk, Russian olive, and screwbean mesquite. Redwoods, white alder, box elder, Oregon ash, and California white oak commonly inhabit the various California floodplains, and in the Pacific Northwest, red alders dominate the floodplain, gradually deferring to bigleaf maple, western hemlock, and sitka spruce.

Moisture-loving vines proliferate in the floodplains, as do ferns, which require a damp environment for reproduction. Other herbaceous plants also thrive, but may be reduced in areas of frequent flooding.

Seldom found far from water, turtlehead (**left**) and the brilliant cardinal flower (**below**) are regular eastern floodplain residents. Both are late summer bloomers.

Some common plants of floodplain forests

(Consult your field guides to determine geographic distribution.)

Trees

green ash
Fremont cottonwood
eastern sycamore
sandbar willow
American elm
black ash
red maple
black maple
tulip poplar
blackgum
water elm
water oak
Nuttall oak
swamp chestnut oak
waterlocust
quaking aspen
blue spruce
redwood
screwbean mesquite
Oregon ash
red alder
western hemlock
eastern cottonwood
black cottonwood
black willow
peachleaf willow
slippery elm
boxelder
silver maple
American beech
overcup oak
pecan
swamp white oak
willow oak
post oak
water hickory
sweetgum
balsam poplar
French tamarisk
Russian olive
white alder
California white oak
bigleaf maple
sitka spruce

Wildflowers and vines

poison ivy
trumpet creeper
bur cucumber
American black currant
spotted jewelweed
turtlehead
trout lily
mayapple
Dutchman's breeches
Spanish moss
river grape
peppervine
Virginia creeper
wax currant
pale jewelweed
cardinal flower
sweetflag
stinging nettle
virgin's bower

ROCKY MOUNTAIN FORESTS

The forests of the Rockies actually consist of four major forest types characterized by the dominant tree species. Unlike eastern forests, the boundaries of the mountainous western forests are defined more by elevation than by latitude. And while species diversity in the Appalachian Mountains decreases with temperature from the base of a mountain to the summit, in the Rockies it increases with precipitation from the lower slopes to the treeline, which is largely determined by temperature.

Ponderosa pine forest

When traveling into the Rocky Mountains, whether from the eastern shortgrass prairies or the sagebrush grasslands of the Great Basin, often the first tall woodlands encountered in the foothills are dominated by ponderosa pine. (Ponderosa pine forests are also prominent in the Black Hills of South Dakota and in the Bighorn Mountains of northeastern Wyoming.) It inhabits the warmer, drier lower slopes of this range, abutting not only grassland, but also sagebrush, pinyon-juniper woodland, and dense, evergreen oak thickets called chaparral.

Trees in the ponderosa pine forest are generally well spaced to avoid competition for moisture and nutrients, conveying an open, airy feeling compared to the thick, lush eastern forests. The understory is frequently reduced or absent, but areas with some moisture support understories and

Above Western serviceberry is a familiar deciduous understory tree in ponderosa pine forests.

Some common plants of ponderosa pine forests

(Check your field guides to determine geographic distribution.)

Trees and shrubs

ponderosa pine	Douglas fir
blue spruce	lodgepole pine
birchleaf spirea	western serviceberry
western chokecherry	mountain snowberry
common juniper	bearberry
black hawthorn	creambush
Pacific ninebark	antelope brush
common snowberry	Colorado pinyon
Rocky Mountain juniper	rubber rabbitbrush
big sagebrush	wax currant
western larch	

Grasses

Arizona fescue	bluebunch wheatgrass
blue grama	Idaho fescue
cheatgrass brome	needle and thread
western needlegrass	western wheatgrass

Left Quaking aspens mingle with ponderosa pines in Custer State Park, South Dakota. Ponderosa pine forests are found in the Black Hills of South Dakota and lower, drier elevations in the Rockies.

Right This view of Moraine Park in Rocky Mountain National Park, Colorado shows widely spaced ponderosa pines on the sunny, dry, south-facing hillside. The cooler, damper north-facing ridge in the middle hosts a Douglas fir woodland, and subalpine forest is visible in the background.

shrub layers that may include western serviceberry, western chokecherry, mountain snowberry, common juniper, bearberry, black hawthorn, creambush, Pacific ninebark and birchleaf spirea. In drier areas, sparse grasses and wildflowers may be the ponderosa pines' only companions.

Douglas fir forest

Higher into the Rockies, the open ponderosa pine merges with the denser Douglas fir forests of midslope. Douglas fir is particularly prominent in the northern Rockies west of the Continental Divide. East of the Divide, cooler temperatures and diminished rainfall conspire against the Douglas fir, and in the southern Rockies, precipitation is also its limiting factor. Douglas fir can form open forests, as the ponderosa pine does, or dense stands.

Douglas fir is a climax species in the Rocky Mountains, wresting dominance from pioneering predecessors such as lodgepole pine or quaking aspen. Often, the dark-green Douglas firs of moist, north-facing slopes and sheltered valleys contrast starkly with yellowish-green, widely-spaced ponderosa pines inhabiting sunnier slopes a short distance away. With water in short supply in many

Some common plants of the Rocky Mountain Douglas fir forest

(Check your field guides to determine geographic distribution.)

Trees and shrubs

Douglas fir	ponderosa pine
grand fir	western larch
white spruce	black spruce
western white pine	lodgepole pine
quaking aspen	Engelmann spruce
subalpine fir	narrowleaf cottonwood
thin-leaf alder	willows
blue spruce	water birch
mountain alder	white fir
Chihuahua pine	Arizona pine
western serviceberry	New Mexican locust
Gambel oak	mountain snowberry
birchleaf spirea	Wood's rose
bearberry	creeping barberry
Nootka rose	

Rocky Mountain regions, it is not surprising that linear micro-forests of water-loving species spring up along streams within the drier coniferous forests. Lining the stream banks are deciduous hardwoods such as Rocky Mountain maple, water birch, western serviceberry, mountain alder, quaking aspen, and narrowleaf cottonwood, as well as the coniferous Rocky Mountain juniper and blue spruce. Bearberry and creeping barberry, two low-growing shrubs, create patches of ground cover throughout.

Lodgepole pine forests

Known to early explorers as Big Piney Woods, large expanses of forest populated by lodgepole pine, a pioneer and fire-climax species, are common in the central and northern Rocky Mountains. At times, they may form climax stands when other conifers are inhibited by local soil composition, temperature, or precipitation. On mountain slopes, lodgepole pine forests are typically located between the Douglas fir forest below and the subalpine forest above.

Lodgepole pines are quite flammable. Fire eliminates the older trees while clearing an area for the next generation of lodgepole pines, which are highly intolerant of shade. Some cones remain on the tree, tightly sealed with resin and ready to open in the intense heat of a forest fire, spreading vast numbers of seeds over the newly prepared seed bed. Lodgepole pines grow straight and tall, hence their common name, and often are so close together that the shaded lower branches die and fall off, creating a "bottle-brush" appearance.

The understory, shrub, and herb layers are quite varied due to the extensive geographic range of this forest. Often the understory contains young Engelmann spruce, Douglas fir, or subalpine fir, shade-tolerant species that eventually assume dominance over the lodgepole pines in the absence of another fire.

Some common plants of lodgepole pine forests

Trees and shrubs

lodgepole pine	Douglas fir
ponderosa pine	subalpine fir
red fir	western larch
white spruce	quaking aspen
western white pine	thimbleberry
western serviceberry	creambush
bearberry	balsam poplar
creeping barberry	

Aspen forests

Aspens love nothing better than a good fire. While the blaze kills most vegetation in the area, the extensive root systems of sun-loving aspens remain alive and vigorously send up new shoots, or suckers, which are genetic clones of the parent plant. For this reason, aspens are sometimes called the phoenix tree, seeming to rise immediately from the ashes of destruction, like their mythical namesake. The hearty root system already in place gives these trees a significant competitive advantage over other species that must start over as seedlings and put down their own meager roots.

Like lodgepole pines, quaking aspens are indicative of disturbed areas, and are very good at colonizing them, but are quite intolerant of shade, and quickly give way to the shade-tolerant species that eventually grow up among them. Even so, they often manage to hold their own along mountain streams, subsisting on the sunlight that reaches through the gap in the coniferous canopy.

Below Aspen forests may be found nearly anywhere in the Rocky Mountains below the subalpine zone, and even there individual quaking aspens may mingle with various conifers. They are especially common along streams and rivers.

Above Many members of the herb layer in Rocky Mountain forests are found in a variety of forest types. Heartleaf arnica, for example, is more likely to be found among aspens, ponderosa pines, or lodgepole pines than with Douglas firs.

Their vast range, the largest of any tree in North America, means that their understories are nearly impossible to characterize because of the geographic overlap with so many other species. Invariably, however, they are more diverse than their counterparts in the shady coniferous forests. In the Rocky Mountains, companion species include western serviceberry, mountain snowberry, bearberry, common chokecherry, common snowberry, assorted willows and alders, black cottonwood, narrowleaf cottonwood, ponderosa pine, spreading snowberry, Douglas fir, lodgepole pine, red fir, Jeffrey pine, white fir, Engelmann spruce, upright snowberry, and subalpine fir.

Some common plants of the aspen forest

Trees and shrubs

mountain alder
limber honeysuckle
bearberry
black cottonwood
peachleaf willow
Bebb willow
curlleaf mountain mahogany
red-osier dogwood
Wood's rose
quaking aspen
mountain snowberry
common snowberry
Douglas fir
Jeffrey pine
red fir
Sitka alder
mooseberry
creeping barberry
narrowleaf cottonwood
Scouler willow
antelope bitterbrush
paper birch
Rocky Mountain maple
Nootka rose
western serviceberry
common chokecherry
ponderosa pine
lodgepole pine
Engelmann spruce
white fir
spreading snowberry
upright snowberry
subalpine fir

Some common wildflowers of Rocky Mountain forests

blue columbine
Nuttall's larkspur
queen's cup
sego lily
arrowleaf balsam boot
Fendler's meadowrue
mule's ears
blue-eyed grass
yellow bell
Engelmann aster
red columbine
mountain bluebell
Canada violet
pussy paws
amber lily
wood nymph
western monkshood
arrowhead groundsel
fringe cups
vase flower
twinflower
Hooker's evening primrose
beargrass
common monkeyflower
mission bells
western starflower
pinedrops
pipsissewa
heartleaf arnica
fireweed
spotted coral root
golden columbine
orange agoseris
bunchberry
elegant camas
Rocky Mountain lily
golden lily
yarrow
western wallflower
giant red paintbrush
false Solomon's seal
silky phacelia
mountain globemallow
towering lousewort

PACIFIC NORTHWEST COASTAL FORESTS

The lushness of the coastal forests stretching from northern California to the Kenai Peninsula contrasts boldly with arid landscapes that, in some instances, are located less than two hundred miles inland. Moderate year-round temperatures and high annual precipitation, both the result of the Pacific Ocean's close proximity, encourage forests of massive trees and thick, luxuriant undergrowth. Because they are isolated by the Pacific Ocean to the west, dry mountains and sagebrush desert to the east, and cold arctic tundra to the north, many of the botanical inhabitants of these woodlands are found nowhere else in the world.

The heavy rainfall is the result of two coincidental environmental factors. Relatively warm westerly winds pick up moisture as they flow over the Pacific's great expanse. Upon reaching land, they are abruptly forced upward by towering coastal mountain ranges. One of the basic principles of meteorology is that warm air can hold more moisture, expressed as relative humidity, than can cold air. The warm air cools as it is forced up the coastal mountains, decreasing its moisture-holding capacity and increasing its relative humidity, and clouds of water vapor begin to form. When the relative humidity reaches 100%, or dew point, precipitation occurs. As the air continues to be forced higher and cooled further, more and more moisture is wrung from the clouds until they crest the highest peaks and begin to flow back downhill. At this point, the process reverses; the air warms as it moves downhill, and precipitation ceases as the relative humidity rapidly drops below 100%, causing a rain shadow, or greatly diminished precipitation, on the leeward side. Desert-like conditions therefore result from the persistent, drying winds flowing down the eastern mountain slopes, only a short distance from temperate rainforests on the western side.

Sitka spruce-western hemlock forests

The Pacific Northwest coastal forests are fragmented into several subtypes, depending upon their locations. From southern Alaska to southern Oregon, western hemlock and Sitka spruce are the dominant trees along the immediate coast and

inland along rivers, and create the lushest understory of any North American forest. Because of the stifling mat of vegetation covering the ground, the conifers depend heavily upon "nurse logs," fallen, decaying timber rich in nutrients, for the germination of their seeds.

Douglas fir forests

Unlike their counterparts in the Rocky Mountains, the Douglas fir forests of the Pacific Northwest are generally not climax communities. They colonize burned areas, but are eventually succeeded by western hemlock in all but the drier areas. Even so, they are associated with the so-called "old-growth" forests of this region, stands of which may exceed 600 years in age. Common understory species in these forests are Pacific madrone, bigleaf maple, and Pacific yew. Other locally common trees are sugar pine, incense cedar, tanoak, canyon live oak, Port Orford cedar, coast live oak, California laurel, and giant chinkapin.

True fir forests

True fir forests, although located at higher elevations and/or more northerly latitudes than either of the first two Pacific Northwest coastal

Left Sitka spruces and western hemlocks dominate the magnificent Cathedral Grove on Vancouver Island, British Columbia. The towering giants of Pacific coastal forests result from moderate temperatures and heavy rainfall, both due to the close proximity of the Pacific Ocean.

Below Old-growth Douglas fir forests have shrunk to a fraction of their former range due to timber companies' continuing harvest of these ancient monarchs.

Some common plants of Pacific Northwest coastal forests

Trees and shrubs

Douglas fir
Sitka spruce
Pacific silver fir
western white pine
bigleaf maple
subalpine fir
Engelmann spruce
lodgepole pine
mountain hemlock
incense cedar
canyon live oak
red alder
evergreen huckleberry
alpine laurel
red huckleberry
devil's club
coast live oak
giant chinkapin
delicious huckleberry
Pacific blackberry
Pacific rhododendron
creambush
western hemlock
noble fir
western red cedar
Pacific madrone
Pacific yew
grand fir
western larch
Alaska cedar
sugar pine
tanoak
Port Orford cedar
black cottonwood
crowberry
wild cranberry
fool's huckleberry
coastal red elderberry
California laurel
thinleaf huckleberry
snow bush
Cascade Oregon grape
salal
vine maple

Wildflowers and ferns

starry Solomon's plume
queen's cup beadlily
vanilla leaf
fireweed
white-flowered hawkweed
evergreen violet
Oregon goldthread
western starflower
beargrass
common monkeyflower
calypso
orange agoseris
Washington lily
pinedrops
Canada violet
queen's cup
Indian pipe
glacier lily
pipsissewa
sword fern
rosy twisted stalk
foamflower
pearly everlasting
snow queen
rattlesnake orchid
yellow skunk cabbage
western monkshood
northern inside-out flower
towering lousewort
Hooker's evening primrose
spotted coral root
tiger lily
blue anemone
bunchberry
fringe cups
phantom orchid
twinflower
oak fern

Left Redwood forests exist only in California and extreme southwestern Oregon, just inland from the coast, but they are worthy of mention because of their uniqueness. They require the deep, well-drained soils of floodplains and river deltas; they also need humid conditions, hence their location in what is known as the fog belt, where dense fogs sweep in regularly from the sea.

Right Relatively few wildflowers thrive in the dim, humid interiors of redwood forests. One of those species that fares very well here is redwood sorrel.

forests discussed, have many species in common with them. The major difference is that here, noble fir and Pacific silver fir dominate in place of the less common western hemlock and Douglas fir. Western red cedar and western white pine are also notable companions of the dominant species. Locally common are subalpine fir, grand fir, Engelmann spruce, western larch, lodgepole pine, Alaska cedar, and mountain hemlock. Thinleaf huckleberry is the characteristic shrub of true fir forests.

Redwood forests

The southernmost of these forests are the redwoods of northern California. They are true giants among trees, and their immensity is mind-boggling, for their trunks may exceed 15 feet in diameter and 300 feet in height. The more ancient specimens are over 2,000 years old, a testament to their extremely thick fireproof and insectproof bark, flood tolerance, and regenerative powers in case of injury.

Only the most shade-tolerant plants grow under the sun-blocking canopy of these mammoths: moisture-loving ferns, mosses, bryophytes, and lichens create a thick green cushion. The high humidity and warm temperatures encourage decomposition, and myriad fungi take advantage of this. Wood fern, sword fern, and chain fern are among the more common species that thrive. The shrub layer, dominated mostly by members of the heath family and quite dense at times, may include Pacific rhododendron, western azalea, evergreen huckleberry, California huckleberry, California hazelnut, vine maple, upright snowberry, and Pacific bayberry. Full-grown conifers of other species may mingle with the redwoods, forming a tall understory; among these may be Douglas fir, western hemlock, Sitka spruce, or grand fir. Tanoak and Pacific madrone, the two most common broadleaf trees in the redwood forest, are occasionally joined by Pacific dogwood, California laurel, Oregon ash, bigleaf maple, red alder, Oregon white oak and white alder.

Some common plants of the redwood forests

Trees and shrubs

redwood	Douglas fir
western hemlock	Sitka spruce
grand fir	tanoak
Pacific madrone	Pacific dogwood
California laurel	Oregon ash
bigleaf maple	red alder
white alder	Oregon white oak
Pacific rhododendron	western azalea
evergreen huckleberry	California huckleberry
California hazelnut	vine maple
upright snowberry	Pacific bayberry
thimbleberry	Pacific ninebark
creambush	coyote bush
salal	salmonberry
western red cedar	Tracy willow
Goodding willow	Pacific willow
heart-leaved willow	red willow

Wildflowers and ferns

redwood sorrel	checkermallow
miners lettuce	Pacific houndstongue
fringecup	milkmaids
yerba de selva	true forget-me-not
northern inside-out flower	fireweed
fringed loosestrife	western trillium
red clintonia	evergreen violet
wood fern	sword fern
chain fern	deer fern

WETLANDS

WHAT ARE WETLANDS?

Habitats in which water saturation is the major environmental influence on soil formation and the accompanying plant and animal life are called wetlands. They are transition zones between aquatic and upland habitats. A high water table occurs at or slightly below ground level, so that the soil is saturated, or the soil is covered with shallow water (up to six feet deep, as defined by the United States Fish and Wildlife Service). Wetlands are characterized by emergent vegetation, plants that protrude above the water's surface, and often, their boundaries are indistinct. In gently rolling terrain, there may be a subtle moisture gradient in which the soil's water content gradually increases as you approach an aquatic habitat such as a lake or pond, and the plant species present are increasingly adapted to wet conditions. In steeper topography, the transition zone may be abrupt, perhaps only a matter of feet, and may not even be properly called a wetland at all. And wetlands in flat terrain can occupy hundreds of thousands of acres and be populated by huge numbers of certain species.

With few exceptions, freshwater wetlands are biologically among the most productive areas on earth, surpassed only by estuaries (the meeting of freshwater and salt-water environments) in the amount of organic material produced per acre. This is not surprising, since water is the solvent in which all of life's processes occur, and a surplus of water should equate to increased productivity. Oceans, however, the ultimate aquatic ecosystems, are actually relatively barren compared to wetlands, because organisms also need the favorable temperatures and light, nutrients, and atmospheric gases that characterize so many wetland areas.

Wetland plant adaptations

Plants and animals depend upon respiration to liberate their stored energy and carry out life's functions. Aerobic respiration (respiration in the presence of oxygen), is most common, but in the absence of oxygen, anaerobic respiration takes place.

A major problem to wetland plants is a dearth of oxygen. While we tend to think of soil as a solid substance, it is actually quite porous, and those pores conduct oxygen and other atmospheric gases to plant roots and soil organisms. Many of us have accidentally killed a houseplant with kindness by overwatering it, flooding the pores of the soil and displacing the needed oxygen. If this did not kill the plant directly, it may have created a favorable environment for soil micro-organisms that attacked the weakened plant.

Lack of oxygen in the soil is a major environmental obstacle in wetlands. Plants that flourish here have had to evolve means to acquire the needed oxygen or do without it. Water lilies, representative of deeper wetland water, are submerged except for their flowers and the upper surface of their leaves. Stomata, the pores generally

Left Once viewed as undesirable places to be drained or filled for development, wetlands are now recognized as crucial to the health of our environment. Wetlands cleanse pollutants from water, recharge groundwater supplies, regulate flooding, and protect shorelines from erosion. Lacustrine wetlands (**right**) occur around lake and pond perimeters and are characterized by emergent vegetation such as water lilies.

located on the undersides of leaves and responsible for gas exchange with the atmosphere, are located on the *tops* of water lily leaves. As well as stomata on their leaves, many wetland trees and shrubs have lenticels, openings on their bark that expand to admit oxygen as needed. Many wetland plants (water lilies, cattails, pickerelweed and arrowhead) have evolved air-filled conduits in their stems which transport oxgyen to their submerged parts. Some wetland plants (black willows and green ash) have adopted anaerobic respiration, converting the potentially toxic byproducts into more benign compounds to be used for growth.

A few wetland plants have turned their watery environment to their advantage in reproduction and seed dispersal. While many are conventionally pollinated by wind or insects, some, such as hornwort and water celery, have developed specialized flowers and water-borne pollen. Quite a few (sedges, green ash, pecan, cranberries) produce buoyant seeds that float some distance from the parent plant before germinating.

Top Wetlands exhibit vegetation gradients determined by water depth or soil saturation. The two dominant plants shown here are pickerelweed and purple loosestrife.

Above Black willows are common trees of wet sites in the eastern U.S., often growing closer to the water's edge than any other tree species.

BOGS

Bogs are unique in their origin and in the vegetation they support. A bog is a relatively stagnant body of water collected in a depression, often partially or completely covered by a floating mat of vegetation that, in many places, is strong enough to support the weight of large mammals. Jumping up and down on the firmer portions of this mat produces the quaking sensation associated with bogs.

Most northern bogs resulted from the last great Ice Age. The mile-high sheet of ice that covered Canada and northern United States gouged depressions in the earth, and debris carried on the ice like a great conveyor belt was deposited in huge piles, called moraine, which sometimes dammed valleys that had been widened and deepened by the glacier. About 10,000 years ago, as the climate warmed, the southern terminus of the glacier began to melt faster than the ice could advance, and it retreated. In its wake, the depressions and blocked valleys filled with water, some from high water tables, others from precipitation or runoff from the melting glaciers, and lakes formed. Sometimes huge chunks of ice broke off as the glacier withdrew. Partly buried in glacial debris, the ice eventually melted, forming a cold, stagnant pool, the perfect prelude to bog formation. Southern boggy areas in the Piedmont and coastal plain originated after the Atlantic Ocean receded and freshwater lakes formed.

Over time, sediments from erosion and organic material began to fill these depressions as the lakes aged. Sedges growing on the banks wove themselves into floating mats that advanced over the perimeter of the lakes, and shortly thereafter, moisture-loving sphagnum moss found these conditions to their liking and thickly colonized the top of the sedge mats. Soon the mats grew thick enough to support shrubs such as Labrador tea and

Below Bogs are unique habitats. Plants that grow here are adapted to very harsh and specific conditions.

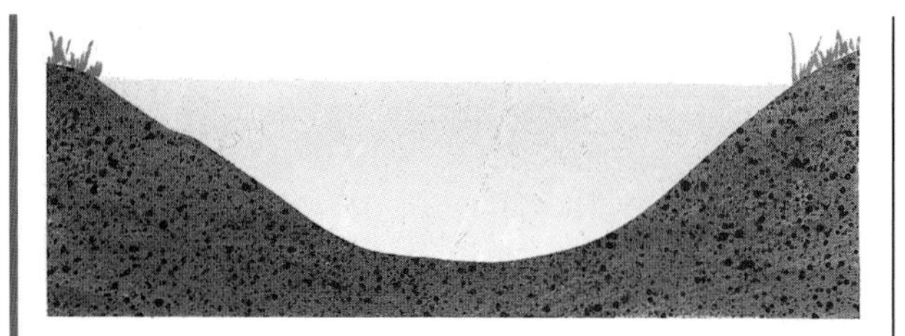

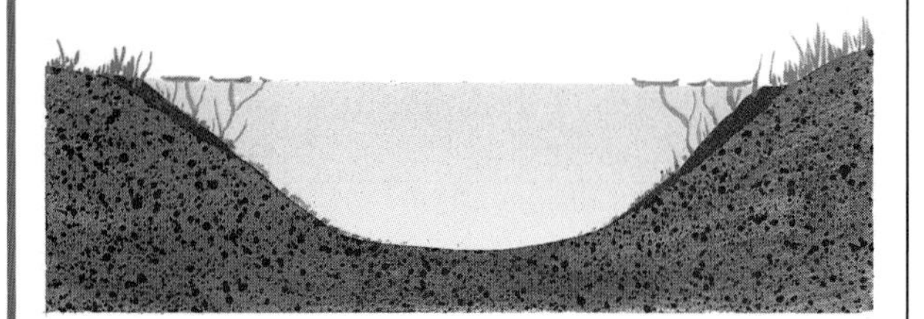

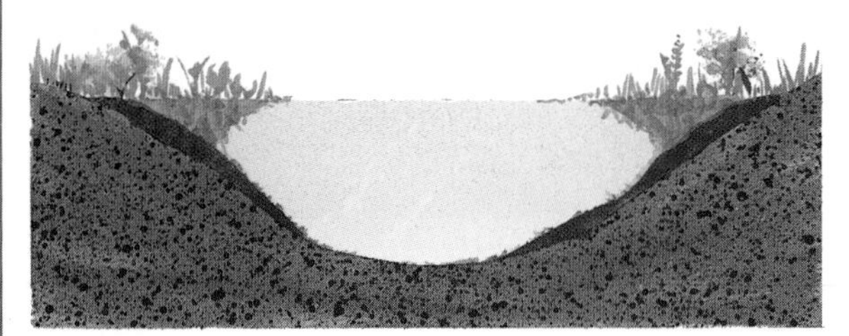

Above There are several types of bogs, the most familiar of which are the quaking bogs. These form as vegetation encroaches over the water's surface from the edge of a lake. These plants become interwoven to form a floating mat of vegetation, which "quakes" when walked upon and eventually may cover the entire lake. As it fills with poorly decomposed organic matter, it becomes known as a peat bog. Because of the underlying wet, acidic conditions it may never succeed completely to the surrounding climax forest.

leatherleaf, whose roots help to bind the mats tightly together.

The water of northern bogs is cold because of its depth, its stagnant nature (no influx of water from a warmer source), the floating mat of vegetation that blocks the sun's warming rays, and the fact that bogs are often located in a depression where pockets of cold air collect. Its temperature, acidic nature imparted by the abundance of sphagnum moss, and low oxygen content, result in decomposition that is incredibly slow, sometimes almost non-existent. The organic material building up the bottom and the vegetation mat sinking under its own increasing weight contain large amounts of poorly decomposed, spongy organic matter, called peat. Where the sinking mat meets the rising bottom and becomes "grounded," black spruce and tamarack trees find enough support to grow. Eventually, the entire bog may fill with compacted peat and become a bog forest, but a bog rarely succeeds to the upland forest of its surroundings. Bog plants are specially adapted to grow in an acidic, nutrient-poor environment, conditions which persist long after the bog has filled in, and which many upland plants cannot tolerate. Also, the water level of even peat-filled bogs remains high, because of the impermeable depression underlying them, resulting in a wet, spongy floor with no true soil.

Surprisingly, plants of these northern wetlands have many characteristics in common with desert plants, since they have to overcome many of the same problems. Many have thick, leathery leaves with a waxy covering to prevent water loss. Some, such as Labrador tea, bog rosemary, and bog laurel, have leaves that curl inward, or a dense wooly covering on their undersides to reduce the flow of dry air over the stomata. Why do wetland plants growing literally on top of the water still need to conserve it? The answer lies with one of the factors responsible for bog formation, the extremely cold water. It is not unusual for the roots of bog plants to be partially frozen in the mat, and therefore non-functioning, as late as July, while the portion of the plant above the mat bakes in the summer sun!

The high productivity of other wetlands, such as marshes, is due to their warmer climates, the efficient nutrient cycling that accompanies rapid decomposition, and their constant recharging with mineral-laden ground water. The lack of such conditions greatly reduces the productivity of bogs. Their cold, acidic nature makes for poor nutrient cycling, and their water level is maintained primarily by mineral-poor precipitation. Two types of carnivorous plants, sundews and pitcher plants, have evolved to cope with these conditions. Sundews secrete jewel-like drops of sticky fluid on the tips of hair-like structures on their leaves. When an insect alights and gets stuck, the hairs draw it down to the leaf surface where enzymes are secreted and digestion takes place. Pitcher plants produce elegantly-veined, vessel-shaped leaves that collect rainwater. Recurved hair-like projections allow insects to crawl into the "pitcher," but pose an obstacle to retreat. Eventually, the insects falter, fall into the accumulated water, and drown. Enzymes secreted by the plant help to digest the food, but most of the breakdown is due to the activity of bacteria.

Bogs are very nutrient-poor environments. Plants like sundews (**above**), and pitcher plants (**below**) have adapted to capture and digest small invertebrates in order to obtain nutrients.

Some common plants of bogs

Trees and shrubs

leatherleaf
large cranberry
bog rosemary
water willow
rhodora
mountain cranberry
highbush blueberry
black spruce
Atlantic white cedar
red maple
eastern hemlock
Labrador tea
sheep laurel
bog laurel
sweet gale
creeping snowberry
white swamp azalea
tamarack
northern white cedar
poison sumac
balsam willow
balsam fir

Wildflowers

grass pink
bog rein orchid
pitcher plants
pink lady's slipper
false Solomon's seal
marsh blue violet
swamp pink
rose pogonia
white fringed orchid
sundews
showy lady's slipper
goldthread
swamp loosestrife

Other herbaceous plants

cotton grasses
sphagnum mosses
bladderworts
sensitive fern
bulrushes

MARSHES AND WET MEADOWS

Marshes are among the wetland systems classified as *palustrine*, a category which includes bogs, swamps, and wet meadows, and accounts for more than 94% of all freshwater wetlands in the lower 48 states. The remaining wetland systems are *lacustrine*, found around lake perimeters; and *riverine*, which occur along rivers and streams.

Most marshes are a mid-stage in the ecological succession of a deeper aquatic habitat. Their longevity as marshes depends on a variety of factors, including size, drainage, and general climate. Some exist in the shallower perimeter of a lake or pond, or in poorly drained depressions along streams and rivers. Others may represent an entire lake in an advanced state of eutrophication, or aging, as sediments and organic matter gradually fill it in. Eventually, this marsh is likely to succeed to a wet meadow, which is commonly found on the upland side of a marsh. As the marsh dries, life becomes impossible for water-loving plants, which give way to those less tolerant of submersion. The wet meadow may ultimately become a dry upland meadow and evolve into the climax community of the area, unless an outside force, such as man, fire, glaciation, or geological activity, intervenes. Or, if periodic flooding continues, it may remain wetland, perhaps succeeding to a shrub or forested swamp in a warmer, drier climate, or to a bog in a cooler, wetter climate.

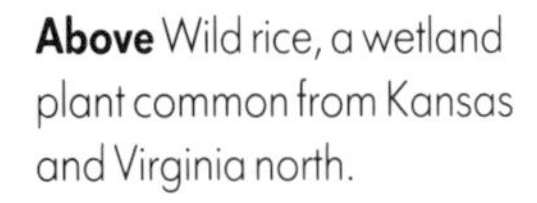

Above Wild rice, a wetland plant common from Kansas and Virginia north.

Above Sedges, grass-like plants of wetlands, have stems that are triangular in cross-section.

Continuously flooded shallow marshes, where water levels range from six to twelve inches during the growing season, are characterized by herbaceous, emergent plants such as cattails, golden club, arrow arum, arrowhead, pickerelweed, bulrushes, spikesedges, smartweeds, bur reeds, and others. Water hyacinth, an exotic South American species, is prolific in southern marshes. Deeper marshes also host reeds and wild rice, as well as submerged species such as water lilies, water celery, bladderworts, and

Left Cattails spread by branching, horizontal rootstocks, often forming dense colonies that dominate marshes. The flowers of this monocot grow in dense spikes, with the male flowers on top. By early summer the stamens are shed, leaving only the club-like mass of female flowers below.

Some common plants of marshes

Grasses, rushes, and sedges

bulrush (a sedge)	spike rush
wild rice	tussock sedge
wool grass	soft rush
giant reed	reed canary grass
rice cut grass	millet
slough grass	foxtail grass
manna grass	white top
maiden cane	blue joint grass
spike grass	giant reed
tule	leafy three-square

Marsh wildflowers

cattails	bur reed
arrowhead	pickerelweed
purple loosestrife	sweet flag
blue flag	yellow iris
Joe-Pye weeds	boneset
yellow pond lily	fragrant water lily
marsh pink	water plantain
arrow arum	golden club
swamp milkweed	fringed loosestrife

Free-floating plants

duckweed	water meal
water hyacinth	

Wet meadow plants

ironweed	Turk's cap lily
yellow-eyed grass	grass pink
spotted jewelweed	marsh marigold
Canada anemone	marsh bellflower
marsh St. Johnswort	swamp candles
arethusa	rose pogonia
fringed gentian	common ladies' tresses
Canada lily	purple-fringed orchid
yellow-fringed orchid	cardinal flower
three-leaved sundew	venus flytrap
three-awn grass	beak rush
marsh fern	sensitive fern
steeplebush	buttonbush
red maple	black ash

Above Arrowhead, obviously named for the shape of its leaves, produces male and female flowers in separate whorls of three. In wetlands, it is usually located between cattails of shallow water and pickerelweed of slightly deeper water.

pondweeds. Free-floating duckweed, the smallest flowering plant, with leaves no bigger than a matchhead, often coats the surface of open water, and microscopic algae, the base of many marsh food chains, abounds.

Some marshes, particularly those in the prairie pothole region of the upper Midwest, dry out completely during seasonal droughts, resulting in an ongoing struggle between upland and wetland plants. In the arid Great Basin region, excessive evaporation leaves behind high concentrations of salts, creating saline marshes that are fed by freshwater sources. The tule marshes of northern California and southern Oregon, renowned for their tremendous congregations of waterfowl during migration, are examples of saline marshes.

Wet meadows either flood seasonally or merely remain saturated. In wet meadows on the upland side of marshes, truly aquatic species yield to moisture-loving grasses, sedges, rushes, and such wetland wildflowers as jewelweed, marsh marigold, grass pink, yellow-eyed grass, ironweed, and Turk's cap lily. The grass-like plants may seem difficult to tell apart. Three-sided sedges are the easiest to distinguish; remember the saying, "Sedges have edges." Round-stemmed grasses and rushes look similar, but rushes are solid stemmed and grasses are hollow stemmed.

Marshes are common throughout North America, especially east of the Rockies. They reach their greatest density in the prairie pothole region of North Dakota, South Dakota, Minnesota, Manitoba, and Saskatchewan; tragically, these are disappearing rapidly due to farmers filling them in and converting them to agricultural usage. Wetlands in general have long been undervalued by most people, who saw them only as wasted space, as barriers to development, to be drained and filled to become "productive." Only recently have people become aware of the vital role that wetlands play in maintaining the health of the world. They soak up flood waters like megasponges and slowly release them again, preventing widespread damage and erosion. They filter pollutants from the water and recharge the underground aquifers that supply much of our drinking water. They are the nurseries for a large portion of the world's food supply. They are undeniably indispensable.

EVERGLADES

Though technically a marsh, the Everglades are so large and its plant population so exotic that it deserves special consideration. It also encompasses numerous "hammocks", raised, forested islands in a sea of grass. These hammocks actually consist of subtropical forests unknown in the United States outside of southern Florida.

The Everglades are fed by a wide, shallow river originating from Lake Okeechobee and flowing in sheets forty miles wide and less than one foot deep.

Above Though more akin to subtropical forests than to wetlands, the hammocks that dot the Everglades are an integral part of this vast wetland. Plants like the gumbo-limbo tree, epiphytes, and the bizarre strangler fig make these a welcome botanical diversion from the sea of sawgrass.

A vast sawgrass "prairie" hides the river from view in many places. Sawgrass, a type of sedge sometimes reaching ten feet in height, spreads by underground rhizomes and, in many places, forms pure stands that tend to exclude other species. Still, many herbaceous plants do make their home among the sawgrass, including white-topped sedge, water pimpernel, salt-and-pepper, blue-eyed grass, bluehearts, wild petunias, yellowtop, horrid thistle, and marsh pinks. Waxmyrtle, a shrub, also makes periodic appearances.

The real wealth of plant life in the Everglades resides in the hardwood hammocks, raised areas one to three feet above their surroundings that are high enough to keep most of the tree roots above the water-soaked soil of the sawgrass prairies. Most of these tropical hardwoods originated in the West Indies and are not found anywhere else in the United States.

A common species and the largest of these trees is the gumbo-limbo, easily recognized by its flaking, coppery bark. Poisonwood is perhaps the most abundant tree. Related to poison ivy and producing similar effects, it is easily identified by the bright yellow margin and midrib of its shiny green leaves. The West Indies mahogany was once abundant but has been extensively harvested. Royal palm, winged soapberry, snowberry, pigeon plum, Jamaica dogwood, and wild coffee are among the tropical species that grow here. More temperate-climate species, such as live oak, red mulberry, and sugarberry, are interspersed with these and other tropical trees.

One of the more unusual inhabitants of tropical hammocks is the strangler fig. After being excreted by a bird onto a tree limb, the undigested seed of the strangler fig germinates and sends aerial roots downward around the trunk of the tree and into the ground, and grows upward into a leafy crown. Upon reaching the ground, the roots expand and fuse, killing the host while the strangler fig remains.

Numerous vines, notably moon flower, inhabit hammock forests, climbing the trees to ensure their own place in the sun above the dense shade of the canopy. On their way up, they pass through a virtual hanging garden of orchids and other epiphytes, the so-called "air plants." Epiphytes are not parasitic; their roots serve as anchors and they utilize various mechanisms to collect precipitation or dew and nutrients naturally leached from the surrounding foliage. Resurrection fern, sword fern, long strap

Some common plants of the Everglades

Trees and shrubs
coastal plain willow
baldcypress
redbay
gumbo-limbo
West Indies mahogany
winged soapberry
pigeon plum
wild coffee
sugarberry
strangler fig
buttonbush
waxmyrtle
pondcypress
sweetbay
poisonwood
royal palm
snowberry
Jamaica dogwood
live oak
red mulberry
southern bayberry
marlberry

Epiphytes
Florida butterfly orchid
clamshell orchid
Spanish moss
wild pine

Wildflowers
water pennywort
arrowhead
common butterwort
water lettuce
grass pink
spider lily
fragrant water lily
duckweed
arrow arum
yellow butterwort
water hyacinth
common cattail
rose pogonia
fire flags
swamp lily
floating hearts

Ferns
wild Boston fern
fan maidenhair
maiden fern
resurrection fern
long strap fern
sword fern

Sedges
sawgrass
white-topped sedge

fern, and Florida butterfly orchid are but a few epiphytic species to be found here.

Islands of woody species not yet developed into hammocks are referred to as heads. They begin as willow heads when coastal plain willows colonize wet depressions during a dry period. The ensuing buildup of organic matter raises the soil to a slightly elevated, less saturated mound that can support various bay species, such as redbay, bull bay, and sweetbay. Cypress domes, so named because of the domed shape of their canopies, develop when baldcypress or pondcypress become established in some shallow ponds.

Left An introduced tropical plant, water hyacinth has been clogging the waterways of southern states, including the Everglades.

Above Water lettuce, a small, floating aquatic plant, often grows so densely as to give the illusion of solid ground.

SWAMPS

The presence of woody trees, shrubs, or both, differentiate swamps from marshes. Northern forested swamps are most prevalent in the Great Lakes states of Minnesota, Wisconsin, and Michigan, while shrub swamps are common both here and in the Piedmont region and coastal plains. Both types of swamps are also present in New England and in isolated regions west of the Mississippi River. Cypress swamps and southern hardwood swamps are the forested wetlands of the Southeast, and are located in lowland areas along rivers and in the coastal plain.

Since swamps often occur in river floodplains, it can be difficult to distinguish them from floodplain forests, especially because there is a great deal of species overlap. The general distinction between the two lies in the proportion of time they are submerged. If the area is inundated from time to time for short periods when the river rises above flood stage, but drains completely when flood waters recede, it is a floodplain forest. If drainage is poor and the soil is covered with water for longer periods, it is a swamp. Obviously, there is a large gray area between the two extremes, and some regions could be classified as either.

Shrub swamps

Shrub swamps may be characterized according to species. Pussy-willow swamps and alder swamps are common across much of the continent. Pussy willows are well known for their fuzzy gray catkins that appear as one of the first sure signs of spring. As these catkins expand and flower, it becomes evident that female and male flowers are produced on separate plants, the bright yellow male flowers easily distinguished from the dull green female flowers. Pussy willows commonly grow in thickets and may reach nine feet in height. They may also include silky dogwood and red-osier dogwood. Beneath their low canopy, many wildflowers, such as bedstraws, tall asters, bluebells, and goldenrods may grow. Alders, close relatives of birch, also produce dense thickets in wetland areas. Buttonbush, an emergent shrub, can thrive in wetter areas than either alders or pussy willows.

A unique type of shrub swamp, called the pocosin (an Algonquin Indian name that literally means "swamp on a hill"), develops in southern

boggy areas. This is the home of small evergreen trees such as redbay, sweetbay, loblolly-bay, and pond pine, as well as shrubs such as waxmyrtle, zenobia, titi, and fetterbush. High water levels and fire help to maintain pocosins and prevent their succession to a higher seral stage.

Northern swamps

Boreal swamps are populated mostly by the same trees that characterize the boreal forest. Farther south, red maple swamps are common, as are those populated by black gum, swamp white oak, American elm, yellow birch, white ash, black ash, and sweetgum. Spicebush is common in the understory of northern hardwood swamps. Cinnamon fern, sensitive fern, marsh marigold, and skunk cabbage are just some of the herb layer.

Skunk cabbage is particularly interesting for its ability to generate heat. One of the earliest signs of spring in the swamp, skunk cabbage flowers consist of a sheath-like spathe that nearly encloses a ball-like cluster of flowers, the spadix. Using the food reserves in its huge roots, skunk cabbage begins to generate heat inside the spathe in late winter, enough to maintain a temperature of 68 degrees Fahrenheit around the spadix even when external temperatures are sub-freezing. This adaptation not only allows skunk cabbage to flower early, but the warmth it produces helps to disseminate the fetid odor that attracts its primary pollinators, carrion flies.

Left Cypress swamps have gained a reputation as dark, brooding places, but they are botanically fascinating.

Below Terrestrial plants have had to make many adaptations to thrive in swamps, such as the aerial "knees" of baldcypress that allow the roots to breathe.

Some common plants of cypress swamps and southern hardwood swamps

Trees and shrubs

baldcypress	black ash
Carolina ash	water tupelo
water elm	buttonbush
redbay	sweetbay
loblolly-bay	black tupelo
Dahoon holly	black willow
honeylocust	overcup oak
swamp chestnut oak	water hickory
Atlantic white cedar	nutmeg hickory
titi	yaupon
waxmyrtle	strangler fig
water oak	black willow
willow oak	sugarberry
sweetgum	water locust
swamp tupelo	ironwood
sweet pepperbush	red buckeye
swamp azalea	flowering dogwood
swamp dogwood	possum haw holly

Ferns

leather fern	resurrection fern
sword fern	strap fern
cinnamon fern	royal fern

Vines

climbing hydrangea	supplejack
American wisteria	Kentucky wisteria
wild grape	peppervine
Virginia creeper	trumpet creeper
crossvine	

Wildflowers

butterfly orchid	clamshell orchid
wild pine	Spanish moss
water lettuce	pickerelweed
arrowhead	fire flags
swamp lily	golden club
crimson pitcher plant	red iris
swamp honeysuckle	spotted touch-me-not
water hyacinth	spider lily
green-fly orchid	switch cane
cardinal flower	butterweed
ladies tresses	rose pogonia
grass pink	water-spider orchid

Southern swamps

Cypress swamps are dark, brooding places in the southeast, characterized by baldcypress trees often enshrouded in Spanish moss. Baldcypress, which is actually not a cypress at all but a member of the redwood family, is a deciduous conifer that produces conspicuous aerial roots, called "knees", to absorb oxygen otherwise unavailable to their submerged roots. They develop swollen, buttressed trunks to help support themselves in the unstable, waterlogged sediments. Air plants such as Spanish moss, bromeliads, and numerous orchids perch on the branches of baldcypress and other trees, obtaining moisture from the rainwater they collect, and nutrients and minerals from dust in the air and from those naturally leached from their host's foliage and bark.

Southern hardwood swamps are regions flooded for less than four months a year. They also contain baldcypress, but are dominated by hardwoods and have a richer herb and fern layer.

Above Skunk cabbage's high rate of respiration in late winter allows it to metabolize energy in the roots, keeping the air surrounding its flowers at about 70°F.

SEASHORES

PLANT ADAPTATIONS TO SALT

The plants of saltwater environments face rigors unknown to those of terrestrial and freshwater habitats. During osmosis, water molecules tend to flow from regions of higher water molecule concentration (fewer dissolved substances such as salts) to areas of lower concentration (more dissolved substances) in an attempt to equalize the osmotic imbalance. In this manner, vascular plants of terrestrial and freshwater habitats are able to absorb water through their roots because the osmotic concentration (of dissolved substances) of cell fluids is naturally higher than that of the water they absorb.

This situation is reversed in saltwater habitats, where the salinity, or saltiness, of seawater is equal to or greater than that of the fluids in the plant cells. Though surrounded by water, saltwater plants actually live in a very "dry" environment. Without special adaptations, not only could plants not absorb the water vital to photosynthesis, but they might actually lose water to the surrounding seawater and would quickly die.

To overcome this challenge, shoreline plants, known as halophytes ("salt-lovers") have evolved numerous adaptations. Those with the most exposure to saltwater maintain an osmotic pressure in their cells many times that of terrestrial and freshwater plants; in other words, their cytoplasm is saltier. Vascular plants have the special ability to selectively concentrate salt in their roots as necessary to ensure a higher salinity than that of the salt water; therefore, the water they absorb is "fresh," since it is primarily the water molecules only which move toward the higher salt concentration.

Any excess salts the plants do absorb must be secreted, lest the plant absorb too much water, rupturing the cells under the tremendous osmotic pressure. To accomplish this, many of these plants have special salt-secreting glands on their foliage. Often, salt crystals or droplets of extremely salty water may be observed on the undersides of leaves; these will be washed away by the next high tide or rainfall. Other plants excrete salt through their roots, and still others accumulate salt in their leaves, which are then shed.

Like desert plants, shoreline plants must also protect themselves against "drought," when saltwater does not reach them, rainfall is lacking, air temperatures and evaporation rates are high, and the sun is beating down. To combat this, they produce a thick, waxy covering on their leaves to minimize water loss through evaporation, just like their desert counterparts. Many are also succulents, storing large amounts of water in their tissues for just such situations.

Shoreline zonation

Just as plants around freshwater habitats grow in distinct zones according to their tolerance for water-saturated soil, plants at the meeting of saltwater and terrestrial or freshwater habitats also grow in distinct vegetation zones. In this case, however, the limiting factors are not only moisture but also salinity. Those with the greatest tolerance for salt will grow closest its source. As the concentration of salt or the chance of encountering salt from high tides or spray decreases, those less tolerant of salt are able to establish themselves. This is particularly evident in the salt marshes associated with estuaries, the areas where saltwater mixes with the freshwater of streams and rivers. In the flat salt marshes of the Atlantic and Gulf coasts, a difference of as little as two inches in elevation can determine two distinct plant communities with differing salt tolerances. The borders of such zones are often quite well defined, but are not always obvious from ground level.

The meeting of land and sea is an extremely demanding environment for plants. Plants of rocky shores (**above**) arrange themselves in distinct zones, while the vegetation of sandy shores (**right**) is concentrated behind dunes well back from the water's edge.

BEACHES AND DUNES

Beaches are a restless and unstable environment, constantly at the mercy of shifting winds and unpredictable wave action. Because of this, the salty conditions, and the searing temperatures as the sand bakes in the sun, little grows on them. It is not until you examine the higher dunes above the reach of the tides that plants grow with any frequency.

Dune formation begins with any irregularity, such as a piece of debris, on the beach that disrupts the onshore winds and causes them to slow and drop grains of sand behind the obstruction. As the mound grows, it causes more disturbances as the wind flows over it, and eventually forms a dune perpendicular to the beach. It is in the shelter of the dune that plants establish themselves. Seeds of American beach grass (sea oats in the Southeast and European beach grass on the Pacific coast) are also deposited with the sand grains. As the plant grows and sends out rhizomes, which in turn sprout new plants, the spreading root systems anchor the sand while the plants further disrupt the air currents and more sand is deposited on the fledgling dunes.

Eventually, these small perpendicular dunes are joined as more sand is deposited between them, forming a secondary dune parallel to the beach. Beach grass advances down the face of the secondary dune, causing new disruptions in the wind flow until a primary dune develops in front of the secondary dune. As the two dunes grow, the hollow between them widens.

Sea oats and short dune grass are also dune colonizers, particularly in the Southeast. Lavender-flowered sea rocket is able to exist on the ocean side of primary dunes by storing water in its succulent leaves and stem. Sea elder, a succulent shrub, and low-profiled seaside spurge also fare well here, as do beach pea, a nitrogen-fixer that fertilizes the soil, and dusty miller. On the back side of primary dunes and on secondary dunes in the East, one may also find glades morning glory, seaside goldenrod, buttonweed, and camphorweed. Marsh hay cordgrass is frequently the dominant grass here, accompanied by sandspur.

In the shelter of the primary dune, vegetation grows more densely, further stabilizing the sand and adding humus, a fundamental component of soil, setting the stage for other plant species to invade. Beach plum, rugosa rose, and bayberry may

Above Sand dunes form a sheltered niche, protected from tides and salt spray, where plants can grow. Plants, in turn, stabilize the dunes to create a fairly long-term habitat.

Below Despite its name, the secondary dune is actually the first of the two to form, beginning with small irregularities disrupting the sand-laden winds.

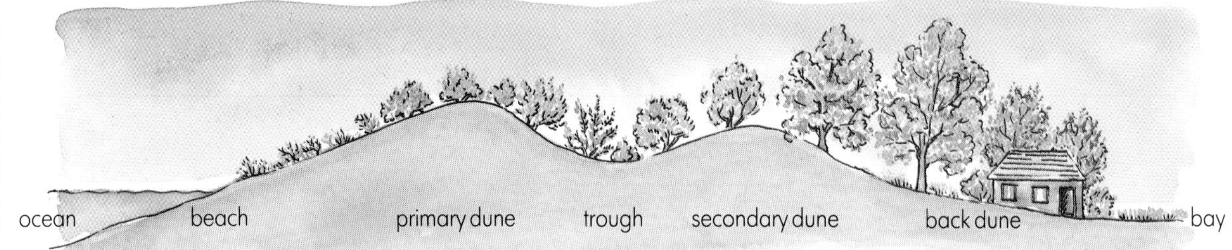

establish themselves in the shelter of the primary dune, providing shade for other plants less tolerant of the blazing sun. Beach heather, poison ivy, and bearberry may help to cover the dunes. Were it not for periodic violent storms, the dune succession would inevitably continue until a climax community was achieved. Indeed, dunes that form farther inland do eventually develop into a climax pine or oak forest, but because moisture drains so quickly in the sandy soil and because these species are such poor nutrient recyclers, they are rarely accompanied by more demanding species.

Pioneering plants on Pacific coast sand dunes include European sea rocket, yellow sand verbena, beach morning glory, silky beach pea, silver beach weed, and American dune grass. A stable plant cover is established by large-headed sedge and seashore bluegrass. Their enhancement of the soil leads to their replacement by such species as seashore lupine, camphor dune tansy, and red fescue. Where protected from the wind, a climax forest characteristic of the area is able to develop.

Yellow sand verbena (**top**), adorned with yellow, trumpet-shaped flowers, is a familiar dune inhabitant of the Pacific coast. Its trailing stem helps keep it ahead of shifting sands. The succulent stems and leaves of sea rocket (**bottom**), a prostrate plant of the Atlantic coast, enable it to retain water under salty conditions, allowing it to colonize unprotected beaches.

Some common plants of sand dunes

Atlantic and Gulf coasts

American beach grass	sea rocket
sea oats	short dune grass
beach pea	dusty miller
seaside goldenrod	sea elder
seaside spurge	glades morning glory
buttonweed	camphorweed
marsh hay cordgrass	sandspur
beach plum	rugosa rose
bayberry	beach heather
poison ivy	bearberry
golden aster	prickly pear cactus
greenbrier	large cranberry
highbush blueberry	sphagnum moss
sheep laurel	pitch pine
red maple	black cherry
eastern red cedar	Japanese honeysuckle
sassafras	American holly

Pacific coast

European beach grass	American dune grass
European sea rocket	yellow sand verbena
beach morning glory	silky beach pea
silver beach weed	large-headed sedge
seashore bluegrass	red fescue
camphor dune tansy	seashore lupine
salal	evergreen huckleberry
ground cone	common ice plant
beach strawberry	beach silvertop
narrow goldenrod	seaside daisy
sea fig	Menzies' wallflower
giant coreopsis	showy Scotch broom
spiny gorse	Douglas fir
Sitka spruce	lodgepole pine
western red cedar	western hemlock

ROCKY SHORES

As difficult as it may seem for plants to survive on coastal sand dunes, it is even tougher for them along the rocky shores of North America, for here there is no protective dune to shelter them and little soil in which to sink roots. Where hardy terrestrial individuals are able to gain a toehold, they must still deal with salt spray and the occasional nor'easter.

Marine organisms inhabiting the intertidal zone constitute the majority of plant life on rocky shores. These must cope with the battering action of waves, desiccation when the tide recedes, salinity, and extreme temperature changes, particularly in the summer when they may spend hours in intense solar radiation, then plunge into frigid waters at the next high tide. Because of their differing tolerances for these variables, plants and animals of the rocky shore distribute themselves in horizontal bands on the rocks.

A variety of lichens colonize the rocks above the highest high tides. *Xanthoria parientina*, a crustose lichen growing within range of wind-driven salt spray, frequently paints the rocks with splashes of bright orange and yellow. Proceeding downward, the first marine-life zone is the black zone, composed of a thick growth of lichens (*Verrucaria* spp.) and blue-green algae of the genus *Calothrix*. To prevent desiccation, these algae are encased in gelatinous capsules; when wet, they are extremely slippery. This zone is often soaked by spray but

Above Irish moss, a leathery alga of variable color and form, is a frequent inhabitant of eastern tidal pools. It is an ingredient in many commercial food products.

Right Few plants can grow in the thick encrustation of barnacles and mussels on rocky shores, but immediately below them lies an extensive zone of rockweeds.

Left Rocky coasts pose special challenges to plants, which grow in specific zones according to their ability to meet those challenges.

rarely inundated except during spring tides, the two highest tides of each month.

Below the black zone lies the upper littoral zone, where conditions are perhaps the most extreme; temperature and salinity may fluctuate drastically in the tide pools of this region. On the Atlantic coast, slender, tubular green algae (*Enteromorpha* spp.) grow in the highest pools. In lower pools are found two red algae: laver (*Porphyra* spp.), growing in wavy sheets or ribbons, and the self-descriptive thread algae (*Bangia* spp.).

Next is the barnacle zone, where the rocks are often completely encrusted with barnacles and mussels. Below that lies the middle littoral zone, consisting of seaweeds. Seaweeds do not have true roots, but secure themselves against the crashing surf by means of rootlike, grasping structures at their base called holdfasts. The species occupying the uppermost part of this zone is spiral rockweed. Immediately below spiral rockweed, on rocks not directly facing the surf, is knotted rockweed, followed by the tough bladder rockweed. Lower still than the rockweeds are Irish moss, dulse, coralline algae, sea lettuce, and an encrusting red algae (*Hildenbrandia* spp.) reaching past the low tide mark. Below the lowest low tides are the kelps.

The upper reaches of high tide on the Pacific coast's rocky shores are also denoted by a black band, and barnacles inhabit the upper littoral zone here as well. Seaward from the barnacles are various species of rockweed plus sea lettuce and sea sack, followed by a red algae, called Turkish towel, and sea palm.

Above Lacking true roots like those of vascular plants, kelps and other colonial algae anchor themselves by means of specialized structures called holdfasts.

Some common plants of rocky shores

Atlantic coast

lichens	blue-green algae
enteromorpha green algae	purple laver
thread algae	spiral rockweed
knotted rockweed	bladder rockweed
Irish moss	dulse
coralline algae	sea lettuce
kelps	encrusting stony red algae

Pacific coast

lichens	blue-green algae
rockweed	little rockweed
sea lettuce	sea sack
Turkish towel	sea palm
winged kelp	sugar wrack
bull kelp	black pine
enteromorpha green algae	coralline algae
nail brush	giant perennial kelp
feather boa	fir needle
encrusting coral	tar spot
surf grass	sea staghorn

SALT MARSHES

Salt marshes form along the coast in areas where the water calms enough to drop its load of suspended sediments. On the low, gently sloping Atlantic and Gulf coasts this can occur behind barrier islands, where wave action is minimal, or along estuaries, where the mouths of rivers fan out, allowing water to slow down while mixing with the salt water of the ocean. On the steeper Pacific coast, salt marshes are limited to narrow strips bordering estuaries. They typically begin as mud flats colonized by algae and eelgrass.

The pioneer species of salt marshes is cordgrass, a very salt-tolerant plant that colonizes the zone between high and low tides. Saltmarsh cordgrass, an eastern species, occurs in two forms. Tall saltmarsh cordgrass, green and up to seven feet tall, inhabits the intertidal zone, while the short, yellow variety establishes itself just above mean high tide. Unlike areas of tall cordgrass, where plant debris is flushed out twice a day at high tide, a mulch is formed by this debris in the short cordgrass, keeping the mud moister and cooler by shading it from the sun. With short cordgrass grow fleshy, translucent-stemmed glassworts, which turn bright red in autumn, and sea lavender, spear scale, and sea blite.

Slightly higher in the marsh is a zone of marsh hay cordgrass, a fine, dense grass that forms tight swirling mats called cowlicks. Spike grass often grows in this same zone. A few more inches in elevation triggers a transformation to black grass, a dark-green plant that turns nearly black in fall. Black grass is covered only by the highest tides of spring and fall, and occasional storm tides. Behind black grass is a shrubby zone of marsh elder and groundsel tree, their further invasion of the marsh inhibited by storm tides. Beyond the reach of tides, the shrub zone may contain bayberry, sea hollyhock, swamp rose mallow, seashore mallow, saltmarsh pink, marsh thistle, glades morning glory, and sea oxeye. The first trees encountered upon leaving eastern salt marshes are usually pitch pines, black oaks, northern red oaks, scrub oaks, and possibly beach plum.

Typical salt marshes are laced with tidal creeks and rivers, drainage channels that flush the marsh with saltwater at high tide and carry it back to the sea at low tide. The exposed banks of tidal creeks contain dense populations of mud algae which are swept seaward with each receding tide. This algae and the tremendous amount of saltmarsh detritus (decomposing organic material), contribute to the enormous productivity of estuaries and intercoastal waterways.

Above Salt marshes exhibit very definite zones of vegetation determined largely by an area's frequency of inundation at high tide. Exposure to salt water is a very rigid limiting factor, and a difference of as little as two inches in elevation can result in two entirely different plant communities.

Pacific cordgrass forms the front line in western salt marshes. Immediately behind it, just above mean high tide, pickleweed reigns, with jaumea and seaside arrowgrass growing in patches among it. Dodder, a bright yellow or orange parasitic plant, often grows in web-like masses among pickleweed.

Right Spike grass is a short, wiry grass which grows at higher elevations in a salt marsh. It often forms dense colonies, especially around the edge of evaporation pools.

Far right Despite the misleading name, saltmarsh bulrush is not a rush, but a sedge with a typical three-sided stem and numerous sharp, pointed leaves. It grows in the intertidal regions of salt marshes.

Above Glassworts, also known as pickleweeds, form mats of prostrate woody stems from which erect, translucent green stems grow. Their succulent stems store large quantities of water necessary for their survival.

Spike grass and marsh gum-plant form the zone behind pickleweed, with Pacific silverweed and various grasses and sedges that cannot stand prolonged inundation with saltwater. Marsh rosemary typically delineates the upland boundary of the marsh. Alkali heath, saltbush, saltmarsh sand spurry, and brass buttons will colonize disturbed sites, as when ocean debris washes up and smothers a patch of pickleweed.

Some common plants of salt marshes

Atlantic and Gulf coasts

saltmarsh cordgrass
sea lavender
sea blite
saltmarsh bulrush
black grass
groundsel tree
sea hollyhock
seashore mallow
marsh thistle
sea oxeye
seaside goldenrod
wigeongrass
glassworts
spearscale
marsh hay cordgrass
spike grass
marsh elder
bayberry
swamp rose mallow
saltmarsh pink
glades morning glory
seaside plantain
saltmarsh aster

Pacific coast

Pacific cordgrass
jaumea
spike grass
alkali heath
saltmarsh sand spurry
marsh rosemary
saltmarsh bulrush
dodder
bear sedge (Alaska)
curlycup gumweed
three-square bulrush
sea milkwort
seaside plantain
pickleweed
seaside arrowgrass
marsh gum-plant
saltbush
brass buttons
Pacific silverweed
tufted hairgrass
creeping alkali grass (Alaska)
eelgrass
redtop
tule

MANGROVE SHORES

Some stretches of the southern Florida coast fit none of the shorelines previously described. These are the mangrove shores, a sort of saltwater swamp. The leggy prop roots of red mangroves allow them to advance into the salt water and to withstand the tides and the abusive waves accompanying tropical storms.

Mangroves are land-builders that extend shorelines and form islands in much the same way that beach grass builds sand dunes. The tangled mangrove roots slow circulating water enough for suspended sediments to settle out and accumulate around the roots. Red mangroves produce the most extensive prop roots and form the outermost zone of advancing mangroves. Its seeds sprout and grow while on the tree, then drop off and float away. They float on their side at first, exposing themselves to the sun so they can photosynthesize their food. As they continue to grow, their center of gravity shifts and the seeds tip vertically in the water, thus increasing their chances of lodging in a suitable location, at which time they undergo phenomenal root growth to quickly stabilize themselves in their new home.

Black mangroves grow nearer to shore in the mud of shallow water previously deposited around the roots of red mangroves. They send up aerial roots similar to baldcypress "knees", to obtain oxygen for their roots buried in the oxygen-poor mud. Black mangroves continue with the process of land-building, and are followed by white mangroves, buttonwood, and finally, a tropical hardwood forest.

Mangroves can be identified by the roots they produce. Red mangroves develop arching prop

Below Because the leggy roots of red mangroves trap suspended sediment, advancing the shoreline as it accumulates, the trees move seaward and are succeeded, in turn, by black mangroves, white mangroves, buttonwood, and tropical hardwoods.

Above Mangrove swamps in North America are found only along the southwestern coast of Florida and in the keys. These red mangroves establish

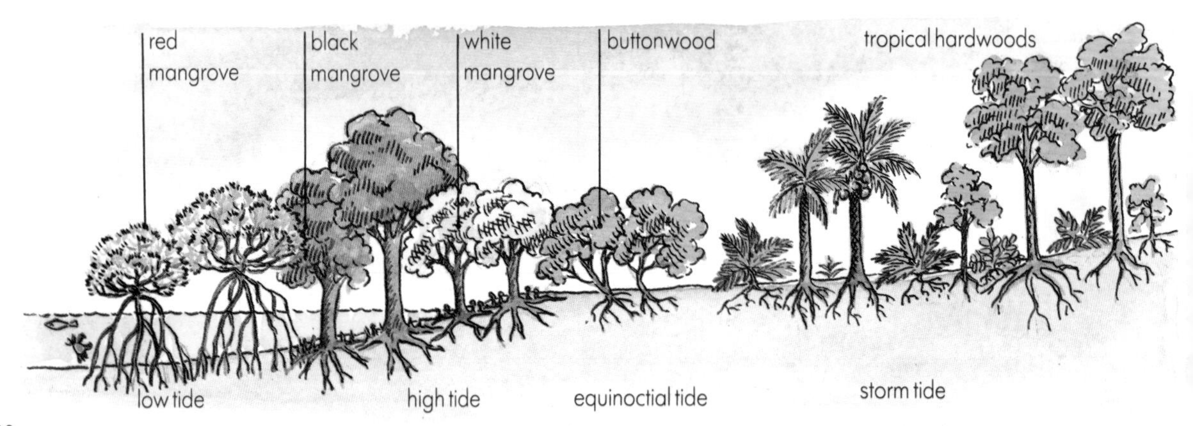

themselves in shallow salt water with virtually no competition from other vascular plants.

roots for support that are also capable of taking in oxygen through lenticels in their bark and shuttling it to the roots buried in anaerobic mud. Black mangroves produce aerial roots, while white mangroves growing on fairly solid ground exhibit neither of these two types. Because mangroves are so well adapted and their environment so demanding, few other species grow with them until the tropical hardwoods or sawgrass take over.

Right Among the tropical hardwoods that succeed mangroves you find species common to the hardwood hammocks of the Everglades, such as sweetbay (bottom). Still farther from the shoreline on higher ground, you will find saw palmetto (top).

Some common plants of mangrove shores and accompanying tropical forests			
red mangrove	paradise tree	slash pine	willow bustic
white mangrove	Bahama lysiloma	saw grass	inkwood
cabbage palmetto	false mastic	glassworts	
saw palmetto	milkbark	sweetbay	
sea grape	Florida nectanra	West Indies mahogany	
sedges	black mangrove	wingleaf soapberry	
strangler fig	buttonwood	gumbo-limbo	

DESERTS
DESERT PLANT ADAPTATIONS

The three most important goals of desert plant adaptations are to conserve water, conserve water, and conserve water! A variation on an old joke, this summation is essentially correct. Instead of not knowing where their next meal is coming from, most desert plants do not know when they will get their next drink. In areas of sparse rainfall, low humidity, and searing heat, plants must be adept at collecting and hoarding every possible molecule of water to survive.

Succulents

The best-known desert plants are the succulents, which include cacti, yuccas, and agaves. Of all the members of the plant kingdom, succulents are the most able to store water. Their leaf or stem cells contain many vacuoles, or storage organelles, which can swell enormously with water when moisture is available to the plant's roots. The stems of many cacti are corrugated, allowing them to expand to accommodate greater volume when more water is being stored. Then, when drought sets in, succulents have a ready supply of water to carry on photosynthesis and maintain turgor, or internal cell pressure, which prevents wilting.

Cacti possess one of the most radical adaptations for preventing moisture loss. Their leaves are reduced to mere spines or bristles because water loss through transpiration is largely proportional to the leaf surface area. The spines are also an effective deterrent to most animals which would otherwise eat the succulent tissue of cacti to obtain moisture and food. Photosynthesis, the major function of leaves, occurs in the outer region of swollen cacti stems.

The stems of cacti and the leaves of other succulents are covered by a leathery or waxy cuticle, an effective barrier to moisture loss. Cacti also have evolved stomata which open only at night, when the air is cooler, to collect and store carbon dioxide in the form of organic acids for the next day's photosynthesis, in contrast with most vascular plants' stomata, which open only during daylight photosynthetic activity. In this way, cacti avoid gaseous exchanges during the heat of the day, which would result in greater moisture loss through transpiration.

One of the more notable features of desert vegetation is that all perennials are rather widely spaced. Succulents have extensive, shallow root systems that quickly gather the sparse precipitation before most of it has a chance to percolate deeper into the ground. Roots of the giant saguaro cactus may cover an area up to 90 feet in diameter. The shallow, spreading roots prevent overcrowding and thus regulate competition for precious resources. The roots of some species actually secrete natural herbicides to keep competing species at bay. "Bare" soil between larger perennials is not necessarily devoid of plant life. Lower plants, such as mosses, lichens, fungi, and algae, may inhabit the sandy

Left Monument Valley, which straddles the eastern border between Utah and Arizona, is a good example of a sagebrush habitat.

Right Desert paintbrush is one of the annuals that splash color beneath the drab greenish gray sagebrush after winter rains.

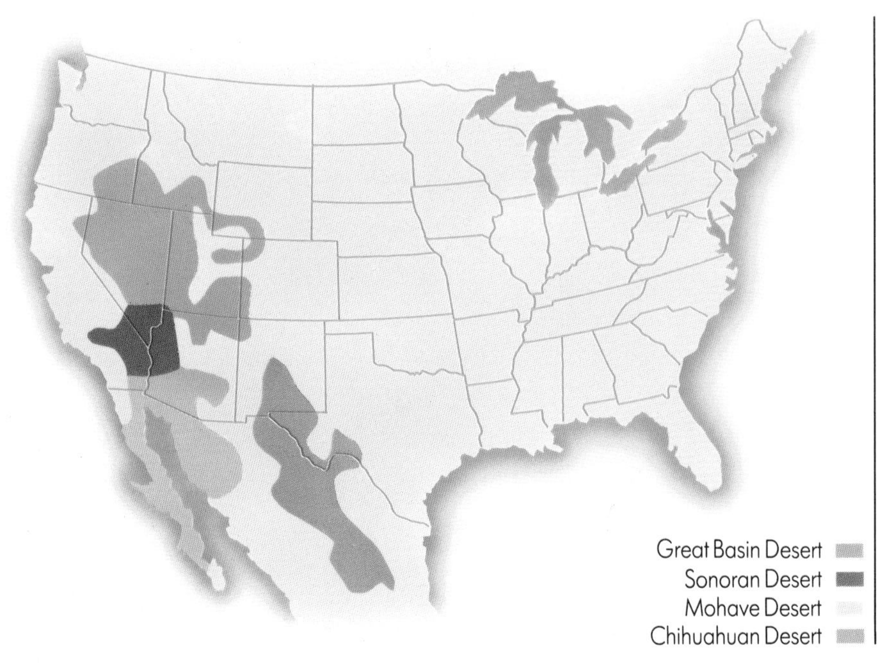

Below Desert plants have had to make some remarkable evolutionary adaptations. Annuals like bladderpods (**below**) have adapted by concentrating their entire life cycle into one brief burst of growth following seasonal rains. In contrast, the stately saguaro (**bottom**), there for the long haul, has combined an extensive shallow root system with a massive water storage capacity.

soil, sometimes forming a stabilizing crust that retards erosion. These plants are able to garner moisture from dew that forms during the night, and can withstand the drying of their tissues until moisture again becomes available.

Desert annuals

Conserving water may also mean becoming dormant during extended droughts. Many deserts do have one or more "rainy" seasons, when they receive a relatively large percentage of their annual rainfall. Desert annuals have adjusted their life cycles to take advantage of the wetter season. Their seeds contain a germination inhibitor that is washed away by a certain minimum rainfall. After these rains, the desert landscape nearly explodes in greenery and flowers as desert annuals rush to complete their entire life cycle within the short period of abundant moisture, for most have no other adaptations to cope with drought.

Desert annuals are divided into two groups: the summer annuals, adapted for life in the intense summer sunlight, and winter annuals hardy enough to cope with cooler temperatures and reduced sunlight. Annuals have protective mechanisms that regulate germination according to temperature and day length as well as precipitation, so that a winter annual is not fooled into germinating during summer rains, only to be faced with searing

summer temperatures for which it is unprepared, and vice versa.

In some desert areas, there is not always a wet season every year, and in such places the seeds of desert annuals have been known to remain dormant for ten years and still sprout with vigor when rain finally falls. In other areas, particularly in the Sonoran Desert, rainy seasons may occur twice a year, in summer and winter. Some annuals sprout during one seasonal rain and enter another dormant period as low-profiled rosettes covered with dense, woolly gray or white hair that drastically cuts the flow of dry air over the leaf surfaces. Their low-rosette stage keeps them out of the moisture-robbing winds until the following rainy season, when they burst into bloom.

Desert shrubs

Desert shrubs grow as low, densely-branched, dome-shaped clumps, reducing their exposure to dry winds. Their leaves are small and often coated with a waxy or resinous substance to retard moisture loss, or with dense, light-colored hair that reduces airflow across the leaf surface and reflects most of the intense solar radiation. Big sagebrush, brittlebush, and some saltbushes can produce two different types of leaves, one for moist periods and another better suited to droughts. In some species, the leaves droop and curl up in response to drought, reducing their exposed surface area. Under more extreme conditions, the deciduous leaves are shed and the shrub enters a dormant period until moisture becomes more abundant.

Creosote bush is a widespread shrub of the hot, southwestern deserts. In the cool desert of the Great Basin, however, shrubs such as sagebrush and saltbushes (which include shadscale, greasewood, hop sage, and winterfat) are dominant, while cacti and other succulents are less prevalent.

Trees

Very few trees can endure the rigors of desert life, but two groups that do are the pinyon pines and junipers. They have found their niche in the upper desert elevations, above the grasslands and sagebrush and below the ponderosa pine forests. These are low trees, usually less than 30 feet tall, with wide, spreading crowns that effectively distribute them to avoid overcrowding, creating an open, grassy savannah. The needles of pinyon pines and scalelike leaves of junipers both reduce

Above One of the few common desert trees, junipers, like this Utah juniper, grow at higher elevations where rainfall is greater.

the surface area they present to drying winds.

The large seeds of pinyon pines, primarily distributed by birds and mammals who either drop them accidentally or store them in caches and forget them, pack enough stored energy to help the new rootlets grow quickly to reach needed water deep in the soil. Other desert plants are also adapted to secure their water from deep in the earth. Certain mesquites, for example, have been found to have roots reaching as deep as 175 feet!

Riparian vegetation

The term "riparian" commonly refers to life associated with river banks, and indeed there are occasional rivers and streams flowing across desert areas. In the desert, however, "riparian" also refers to washes and other drainage areas that may be dry at some times and host intermittent streams at others. The runoff from paved roads may simulate wash conditions at the roadside, creating an artificial riparian habitat. Even shallow drainages less than two inches deep can conduct enough water during rains to make a difference to plants living there; burrobush and smoke tree are common in these temporary drainage channels. Riparian vegetation has adapted to desert life by growing in these areas of relatively abundant moisture. They are often quite different from the plants of the surrounding desert.

Along the permanent desert waterways there are typically two terraces. The lower terrace is covered by ordinary flood waters, while the upper terrace is inundated only during the more extreme floods. The lower terrace is therefore wetter and conducive to the growth of different species than the upper

Left Riparian vegetation grows along desert streams and washes where soils are moister. Cottonwoods grow where there is the greatest amount of moisture, while other riparian species that require somewhat less moisture grow farther from the drainage's center.

terrace. Various willows and cottonwoods grow closest to the water. Farther back on the first terrace arrow weed can often be found, followed by quail bush and screwbean mesquite. Other mesquites inhabit the upper terrace along with tamarisk, a European import, desert broom, seep willow, desert willow, Apache plume, acacias, and others.

Great Basin Desert

Of the four great North American deserts, the Great Basin Desert is the only one classified as a "cold" desert, meaning that it receives most of its precipitation during winter in the form of snow. It is the most northerly of the four deserts, and also has the highest average elevation. Geographically, it overlaps the intermountain grasslands, which occur at slightly higher elevations than does the true desert of this region.

Shrubs dominate the landscape here, and cacti, while present, consist of fewer species and play a less significant role than they do in the hot deserts. Absent is the multitude of annual species that cause the hot deserts to periodically erupt in color. Also absent are the riparian species that line the drainages of hot deserts; having evolved in a warm moist climate, they require both the moisture of the waterways and the warmer temperatures found in the lower elevations of southern deserts.

Sagebrushes, led by big sagebrush, and saltbushes are the dominant shrubs. Grasses commonly accompany sagebrush, particularly at higher elevations. Mormon tea is another shrub often associated with sagebrush, and the region's primary cactus species, plains pricklypear, also mingles with these two species. Common herbaceous species found among sagebrush

Some common plants of the Great Basin Desert

Trees and shrubs

big sagebrush
black sagebrush
green rabbitbrush
greasewood
Mormon tea
iodinebush
snakeweed
turpentine broom
antelope bitterbrush
Wood's rose
tamarisk
Colorado pinyon
cherrystone juniper
hop sage
sand sagebrush
shadscale
winter fat
littleleaf horsebrush
spiny hopsage
rubber rabbitbrush
blackbrush
desert peach
four-wing saltbush
curlleaf cerocarpus
singleleaf pinyon
Utah juniper
cliff rose

Wildflowers

alkali sacaton
pickleweed
Indian paintbrush
lupines
hood phlox
tumbleweed
bur buttercup
clasping pepperweed
blanket flower
hairy golden aster
scarlet gilia
Gunnison mariposa
goldenrods
plains pricklypear
claret cup cactus
Nuttall's larkspur
desert four o'clock
sulphur flower
cushion cactus
desert trumpet
rushes
sego lily
desert paintbrush
scarlet globemallow
longleaf phlox
halogeton
tansy mustard
arrowleaf balsamroot
desert dandelion
locoweeds
mule's ears
meadow death camas
yarrow
Simpson's hedgehog cactus
yellow bee plant
golden prince's plume
desert candle
threadleaf groundsel
tahoka daisy

Grasses

bluebunch wheatgrass
bottlebrush squirreltail
giant wildrye
needle-and-thread
western wheatgrass
red three-awn
Idaho fescue
cheatgrass brome
Indian ricegrass
needlegrass
foxtail barley
secunda bluegrass
prairie junegrass
Nevada bluegrass

Above The Great Basin is characterized by an abundance of shrubs. Mormon tea (**top**) is very common, along with sagebrush, Fewer species of annuals grow here than in other North American deserts, but two typical residents are sego lily (**left**) and Nuttall's larkspur (**right**)

include lupines, globemallows, paintbrushes, locoweeds, phloxes, and sego lily. Saltbushes, like shadscale, winterfat, greasewood, and hop sage, are common in alkaline soils at lower elevations, with shadscale dominating. Iodinebush, saltgrass, rushes, alkali sacaton, and pickleweeds occur on the more saline sites.

In areas disturbed by fire, big sagebrush may be temporarily displaced by snakeweeds, rabbitbrushes, and horsebrushes. Also found on disturbed sites are tumble weed, halogeton, bur buttercup, tansy mustard, and clasping pepperweed. An exotic grass introduced from Asia in the 1870s, cheatgrass brome has displaced much of the native grasses, such as bluebunch wheatgrass. Since cheatgrass brome produces much more fuel (leaves and stems) than the native species, fires that occur are hotter and more frequent than before, which has led to the decline of big sagebrush in some areas.

To the south, where the Great Basin Desert merges with the Mohave Desert, blackbrush predominates, often forming pure stands. This community may also contain turpentine broom, desert peach, Mormon tea, sagebrushes, and saltbushes.

Mohave Desert

The smallest of the North American deserts, the Mohave Desert is slightly higher in average elevation than the neighboring Sonoran Desert, even though it includes Death Valley at 282 feet below sea level. It lies within the basin and range province that constitutes the Great Basin, and some argue that it is really part of the Great Basin Desert, but botanically it is separate.

This desert is dominated by widely spaced, low shrubs. Creosote bush is easily the most prominent plant in the hot deserts of North America, reaching its northernmost limits in the Mohave Desert. Readily recognized by its yellowish-green leaves with their resinous coating and strong resinous odor, and its numerous yellow flowers, creosote bush is extremely efficient at obtaining water from the soil; so much so, in fact, that its own seedlings cannot grow nearby. It frequently reproduces by sending out runners that sprout upon reaching a suitable location. Other shrubs likely to be encountered are Parry saltbush, woolly bur sage, Mohave sage, white bur sage, desert holly, cattle spinach, brittlebush, spiny mendora, goldenhead, blackbrush, and others. In dry washes one is likely

Above The presence of creosote bush defines the hot deserts of North America. The resinous coating of its pungent leaves serves to retard the loss of water vapor.

Some common plants of the Mohave Desert

Trees and shrubs

creosote bush	Parry saltbush
woolly bur sage	white bur sage
Mohave sage	desert holly
cattle spinach	brittlebush
spiny mendora	goldenhead
blackbrush	bladderpod
honey mesquite	screwbean mesquite
Fremont cottonwood	tamarisk
arrow weed	four-wing saltbush
desert willow	seep willow
smoke tree	Anderson lycium
burrobush	catclaw

Wildflowers

California poppy	filaree
fiddleneck	tidy tips
desert dandelion	pincushion flower
owl clover	notch-leaved phacelia
pygmy lupine	birdcage evening primrose
yellow bee plant	Mohave aster
Mohave desert star	Death Valley sage
desert candle	desert chicory
desert trumpet	desert sand verbena
desert four o'clock	desert globemallow
rattlesnake weed	golden prince's plume
western jimsonweed	Canterbury bells
Arizona lupine	spreading fleabane
desert five-spot	desert marigold
desert gold	scalloped phacelia
diamond cholla	barrel cactus
beavertail cactus	claret cup cactus
plains pricklypear	Joshua tree
blue yucca	Mohave yucca

Grasses

big galleta	bush muhly
fluffgrass	Indian ricegrass
gramas	

to find burrobush, catclaw, and honey mesquite.

The Mohave Desert is the site of some of the most awesome displays of desert annuals following favorable winter rains. California poppy, fiddleneck, and filaree may cover acres with uniform color as a single species forms pure stands.

Among the more than 250 species of annuals lending their colors to the seasonal displays are tidy tips, desert dandelion, pin-cushion flowers, owl clover, notch-leaved phacelia, and many others. Eighty percent of annuals found in the Mohave Desert do not exist anywhere else.

Succulents, symbolized by Joshua tree, the largest of the yuccas, come into their own in the Mohave Desert. Blue yucca and Mohave yucca are

Below Joshua tree is indigenous to the Mohave Desert. Not really a tree but a species of yucca, it may grow densely, sometimes forming "forests" within its range.

The Mohave Desert primarily hosts winter annuals; one of the most prolific is California poppy (**left**), the state flower of its namesake, which blooms so profusely that it often turns entire hillsides orange. Desert plume (**above**) produces tall, slender wands of yellow flowers.

also common here. Needless to say, cacti are abundant, though many are low-growing.

Sonoran Desert

No other plant better symbolizes the Sonoran Desert than the stately saguaro. Towering as much as 50 feet above the desert, it is the state flower of Arizona where it and all other cacti are protected by law. The saguaro usually occurs on upper desert slopes together with barrel cactus, ocotillo, golden cholla, teddybear cholla, hedgehog cactus, desert agave, and small trees, such as blue palo verde and desert ironwood. Where the Sonoran Desert merges with the grasslands fringing its upper elevations, desert spoon, with its five- to eight-foot spikes of yellow flowers, thrives.

The valley floors are often dominated by creosote bush and white bur sage or bur sage, as are those in the Mohave Desert. In areas with sandier soil, creosote bush is replaced by Mormon tea and indigo bush. Representative of the cacti at lower elevations are various chollas as well as beavertail cactus. The valley floors and lower slopes are criss-crossed by drainages which harbor just enough moisture to support such riparian species as burrobush, graythorn, and lycium.

In the United States, the eastern portion of the Sonoran Desert, known as the Arizona uplands, is the most scenic and botanically diverse. All species of cacti native to the Sonoran Desert are plentiful. In addition to cacti, foothills palo verde grows everywhere except on the valley floors. Desert ironwood, a small tree, is also characteristic of the Arizona uplands. Noticeable shrubs of this area include whitethorn acacia, creosote bush, bur sage, desert buckwheat, ratany, paperflower, brittlebush, jojoba, fairy duster, and limber bush, with flexible, reddish branches carrying attractive bright green, heart-shaped leaves. Along washes and roadsides, desert willow, desert broom, and burrobush do quite well.

The Sonoran Desert receives the most precipitation of the four North American deserts, and it also receives both late summer and winter rains, so that it supports summer and winter annuals. Colorful spring wildflowers following winter rains include Parry's penstemon, gold poppies, purple mat, Coulter's lupine, chia, owl clover, and pincushion flowers.

Nothing symbolizes the Sonoran Desert better than the towering saguaro cactus (**right**) which may reach 50 feet in height and 200 years in age. Its pleated stems expand in girth to store great quantities of water when precipitation is available. The ocotillo (**left**) is a resident cactus of both the Sonoran and Chihuahuan Deserts.

Some common plants of the Sonoran Desert

Trees and shrubs

bur sage
indigo bush
graythorn
blue palo verde
desert ironwood
whitethorn acacia
desert buckwheat
paperflower
jojoba
limber bush
desert broom
arrow weed
elephant tree
honey mesquite
smoke tree
snakeweed
triangle-leaf bursage
California fan palm
velvet mesquite
white bur sage
Mormon tea
Anderson lycium
foothills palo verde
desert hackberry
creosote bush
ratany
brittlebush
fairy duster
burrobush
desert willow
desert holly
Gregg catclaw
screwbean mesquite
shadscale
tamarisk
chuparosa
Fremont cottonwood
desert spoon

Succulents

beavertail cactus
barrel cactus
golden cholla
jumping cholla
saguaro
ocotillo
teddybear cholla
cane cholla
buckthorn cholla
organpipe cactus
prickly pear cactus
fishhook cactus
night-blooming cereus
soaptree yucca
Parry's century plant
hedgehog cactus
fishhook barrel cactus
desert Christmas cactus
pincushion cactus
Mohave yucca
dudleya
rainbow cactus

Wildflowers

gold poppy
purple mat
Mohave desert star
desert dandelion
wild heliotrope
pincushion flower
birdcage evening primrose
desert four o'clock
desert globemallow
desert sand verbena
yellow bee plant
desert candle
desert trumpet
spectacle pod
desert sunflower
common ice plant
Parry's penstemon
Coulter's lupine
fiddleneck
owl clover
chia
yellow twining snapdragon
Arizona jewel-flower
desert five-spot
desert marigold
western peppergrass
Indian blanket
desert chicory
desert velvet
Mohave aster
yellow cryptantha
Coulter's globemallow

Above With both summer and winter rains, the Sonoran Desert hosts the most species of annuals and cacti of the four deserts. One common Sonoran cactus is beavertail.

Chihuahuan Desert

As in the other hot deserts of North America, creosote bush dominates the lower elevations of the Chihuahuan Desert. Tarbush is often codominant with creosote bush, and mariola, with its pungent foliage, is a frequent companion of both. In some areas, creosote bush is accompanied by whitethorn acacia and sandpaper bush. Crucifixion thorn, buffalo gourd, and Mormon tea likewise frequent the Chihuahuan lowlands, along with herbaceous species such as desert marigold, little golden zinnia, desert zinnia, white horsenettle, and fluffgrass.

Precipitation occurs mostly during the summer, resulting in a preponderance of summer annuals in the valleys, but sufficient winter precipitation occurs in some areas to support winter annuals as well. At slightly higher elevations, the increased annual precipitation brings a corresponding increase in succulents, particularly yuccas and

Above Among the many summer annuals of the Chihuahuan Desert is the little golden zinnia, a member of the sunflower family that grows in low, round clumps.

agaves. Soaptree yucca, banana yucca with its stiff blue-gray leaves, false agave, and lechuguilla, an agave sometimes mistaken for a yucca, are a few that increase in frequency on the lower slopes. Somewhat higher, ocotillo and the shrubs Texas silverleaf and ratany become more common.

Higher still, the desert begins to merge with grassland. Yuccas and agaves, such as Harvard agave and Parry's century plant, so named for the many years that elapse before the mature plant flowers, mingle with two grass-like members of the lily family, beargrass (in New Mexico) and sacahuista (in Texas), with their creamy, club-shaped flowers. Desert spoon, another member of the lily family, produces an elongated spike with thousands of yellow flowers, and is also found in this zone.

Cacti abound throughout the Chihuahuan Desert. Along with such colorfully descriptive names as horse crippler and Turk's head, they may include prickly pear, Texas prickly pear, claret cup cactus, cushion cactus, night-blooming cereus, rainbow cactus, tree cholla, desert Christmas cactus, and clavellina, a low cactus that may form mats up to 10 feet in diameter.

Washes in the Chihuahuan Desert host such riparian species as graythorn, Apache plume, catclaw, and little leaf desert sumac. The lowest depressions are home to four-wing saltbush, burrograss, and tobosa, a coarse grass. Disturbed ground supports snakeweed, tumbleweed, and Russian thistle.

Some common plants of the Chihuahuan Desert

Trees and shrubs

creosote bush
mariola
sandpaper bush
buffalo gourd
Texas silverleaf
graythorn
little leaf desert sumac
desert spoon
Mexican palo verde
screwbean mesquite
tamarisk
tarbush
whitethorn acacia
crucifixion thorn
Mormon tea
ratany
catclaw
honey mesquite
four-wing saltbush
rabbit brush
snakeweed
arrow weed

Succulents

Torrey yucca
ocotillo
false agave
Harvard agave
horse crippler
prickly pear cactus
plains prickly pear
cushion cactus
rainbow cactus
desert Christmas cactus
soaptree yucca
lechuguilla
banana yucca
Parry's century plant
Turk's head
Texas prickly pear
claret cup cactus
night-blooming cereus
tree cholla
clavellina

Wildflowers

bear grass
Apache plume
desert marigold
desert four o'clock
desert poppy
spectacle pod
filaree
little golden zinnia
tahoka daisy
Chihuahua flax
Indian blanket
jackass clover
devil's claw
sacahuista
white horsenettle
desert zinnia
desert gold
pale trumpets
trailing four o'clock
Fendler's bladderpod
Arizona jewel flower
Coulter's lupine
spreading fleabane
golden prince's plume
western peppergrass
desert tobacco

Grasses

Indian ricegrass
burrograss
fluffgrass
tobosa

HIGH COUNTRY

WHAT IS TUNDRA?

Tundra is the term for the colder terrestrial regions where temperature is the limiting factor suppressing tree growth. In North America there are two types of tundra that are different botanically as well as geographically. Arctic tundra is that great treeless expanse north of the boreal forest. Here, the active photosynthetic period of plants is only about three months a year, but due to the long days, the plants photosynthesize food around the clock, so they can reproduce and store enough to last through the long arctic winter. Several feet below the soil's surface, permafrost, permanently frozen subsoil, prohibits the development of tree roots. Permafrost, the severe climate, and the very slow nutrient recycling due to the effect of cold on decomposers, combine to make tree growth impossible.

Alpine tundra, found on mountain summits, is the result of elevation instead of latitude. Though the winters here are shorter, the climate is in some ways more severe than that of the arctic tundra. High winds constantly buffet the landscape, and the thin atmosphere at higher elevations screens out less of the damaging ultraviolet radiation from the sun. Primary succession, the development of soil and plant communities from bare rock, is remarkably slow due to retarded plant activity, and the little soil that does form is subjected to erosion by the strong winds and large volumes of precipitation that accompany high elevations. Therefore, bare rock is a common component of the alpine tundra landscape, colonized only by lichens. In the most exposed locations, even lichens may not grow. Other plants of the alpine tundra are largely restricted to the spaces between rocks where a small amount of soil has accumulated.

Alpine plant adaptations

Despite its heavy annual precipitation, alpine areas are, in effect, a type of desert. Because there is little soil to absorb moisture, much of the water from rainfall and melting snow is unavailable to plants, lost as runoff to mountain streams. In addition, the constant winds, while frequently bearing moisture as they sweep up the mountain slopes and cool past their dew point, nevertheless can rob moisture from exposed plants faster than they can absorb it, so alpine plant adaptations are quite similar to those of desert plants.

Aside from rock-colonizing lichens, most alpine tundra plants grow as dense mats or dome-shaped cushions. In summer, these shapes keep the plants low and out of drying winds while allowing them to present a respectable surface area to the sun for photosynthesis. Their compact shapes also keep them warmer by restricting the access of cold air to the plant's interior, sometimes resulting in temperatures 20 degrees warmer than the surrounding air. Many species grow primarily on the leeward side of rocks or other obstructions, which also helps to keep them warm and moist. In

One most accessible alpine tundra areas is located along Trail Ridge Road in Rocky Mountain National Park, Colorado (**left**). Alpine tundra plants exhibit many of the same adaptations as desert plants; both habitats experience high evaporation rates, and alpine areas have little water-retaining soil. Also adapted to survive low temperatures, many, like moss pink (**right**) grow in low, dense clumps that conserve warmth.

Left Sheltered areas above timberline, such as glacial cirques where soil has had an opportunity to accumulate, support sedges and sod-forming grasses as well as a variety of wildflowers and low-growing shrubs.

the winter, these same shapes allow the plants to remain under a protective blanket of snow, quite warm when compared to the air temperature and wind chill, and out of destructive winds. Those areas that are windswept and barren in winter are also likely to support the fewest plants, and those that do survive are stunted.

Other adaptations common to desert and alpine plants help conserve water and/or reduce exposure to solar radiation. Narrow leaves, dense foliage, and thick cuticles (surface tissue) are common characteristics that reduce evaporation. Some alpine plants store water in waxy, succulent leaves and stems. Leaves of many species also have a dense, woolly covering of light-colored hairs that restrict the flow of desiccating air over their surface, insulate the leaves, and reflect some of the harmful solar radiation bombarding them.

The small size of alpine plants is itself an adaptation to the harsh environment, for the plants use their energy to survive, rather than expending it on large, showy growth. All alpine plants are perennials, another energy-conservation measure which allows them to spend winter in a state of readiness for the unpredictable start of their short growing season. They flower immediately after the snows melt, allowing time for their seeds to be dispersed, germinate, and establish themselves before the next winter starts. One seeming extravagance is the production of large, showy blossoms that often dwarf the rest of the plant, but this is crucial to reproduction, for in such a barren landscape, small flowers could easily be overlooked by bees and other pollinating insects.

Alpine tundra

While we commonly think of alpine tundra as a habitat of the high Rocky Mountain, Sierra-Nevada, and Cascade peaks, there are many isolated examples in the Appalachian Mountains of the northeastern United States as well. Perhaps the best and most accessible alpine tundra flora in the east is via the Auto Road up Mt. Washington in the White Mountain National Forest, New Hampshire. The Alpine Garden Trail, beginning about one mile from the summit, goes through alpine habitat hosting more than one hundred plant species, many of which are relatively rare in the east. Heath shrubs, abundant due to the cold climate, include bog bilberry, mountain cranberry, alpine bearberry, Lapland rosebay, and alpine azalea, moss plant, and mountain heath. Other species likely to be encountered are diapensia, alpine goldenrod, dwarf yellow cinquefoil, alpine bluet, Mt. Washington avens, Greenland sandwort, and wine-leaved cinquefoil. Lichens occur regularly on rock surfaces.

Timberline (the imaginary line beyond which trees do not grow because of cold) in the west is directly related to latitude. In the southern Rockies, tundra does not occur below 12,500 feet, while in south-central Alaska timberline may occur as low as 1,500 feet. One of the more accessible regions of alpine tundra in the west occurs along Trail Ridge Road in Rocky Mountain National Park, Estes Park, Colorado. Following an ancient trail used by the Ute and Arapaho Indians of the area, this road winds its way up to an elevation of 12,183 feet while transversing the major mountain range of the park. Numerous pull-offs and several paved trails, which restrict foot traffic and protect the fragile tundra environment, allow easy access to the alpine ecosystem.

The similarity of their environments means that alpine tundra plants often have a large geographical range; alpine tundra in New Mexico is, after all, much the same as alpine tundra in Alberta. Plant communities within a small area may differ markedly, however, depending upon their exact microhabitats. For instance, plants found on gravel flats and scree slopes are fully exposed to the elements. Those not buried by winter snow tend to be very compact; examples

Alpine wallflower (**top**) grows as a low tuft of basal leaves, keeping out of the bitter, drying winds of alpine tundra. The succulent stems and leaves of king's crown (**above**) store water as a hedge against desiccation.

Some common plants of alpine tundra and meadows

West

mountain sorrel	alpine avens
tufted phlox	greenleaf chiming bells
moss campion	alpine wallflower
fairy primrose	snowball saxifrage
dwarf clover	alpine sunflower
king's crown	arctic gentian
yellow monkeyflower	elephant heads
little elephant heads	dwarf lewisia
eriogonums	sandworts
alpine everlasting	spreading phlox
alpine lupine	alpine columbine
Drummond's cinquefoil	penstemons
club-moss ivesia	primrose monkeyflower
alpine buttercup	sibbaldia
alpine paintbrush	dwarf knotweed
American bistort	granite gilia
goldenbush	whitesquaw
red mountain heather	white mountain heather
shrubby cinquefoil	alpine laurel
alpine willow	pussypaws
glacier lily	beargrass
alpine prickly currant	ballhead sandwort

East

bog bilberry	mountain cranberry
alpine bearberry	Lapland rosebay
alpine azalea	moss plant
mountain heath	diapensia
alpine goldenrod	dwarf yellow cinquefoil
alpine bluet	Mt. Washington avens
Greenland sandwort	wine-leaved cinquefoil

include sandworts, phloxes, eriogonums, spring locoweed, dense-leaved draba, and alpine fescue, a bunchgrass. A similar habitat that does receive snow cover may contain alpine paintbrush, alpine lupine, or club-moss ivesia. By contrast, the relative shelter of boulder fields, talus slopes (piles of rock fragments at the foot of a cliff), and rock crevices is ideal for shrubs such as granite gilia, whitestem goldenbush, and whitesquaw, plus herbaceous species too large to survive elsewhere on the tundra.

Western alpine areas differ from eastern ones in sheer size, and contain more sheltered areas where soil accumulates to form alpine meadows. Such meadows are usually dominated by sod-forming grasses and sedges, particularly hairgrass. Herbaceous wildflowers common to well drained meadows include Lyall's lupine, alpine everlasting, club-moss ivesia, dwarf lewisia, Drummond's cinquefoil, alpine goldenrod, and Sierra daisy. Shrubs are not abundant in alpine meadows, but miniature thickets of various dwarf willow species may be found together with alpine laurel, dwarf huckleberry, white mountain heather, and red mountain heather. Wetter meadows below melting snowbanks or near alpine lakes and streams will host species such as sibbaldia, alpine shooting star, elephant's head, little elephant's head, primrose monkeyflower, alpine buttercup, dwarf knotweed, Anderson's alpine aster, alpine goldenrod, alpine paintbrush and rosy stonecrop. Alpine meadows are fairly sheltered relative to the rocky plant communities found at the summits, and many species found in alpine meadows are also common to subalpine meadows below timberline.

Above Subalpine forests, the transition zones between alpine tundra and montane forests, vary in their species composition with geographic location. Indicator tree species in the Rocky Mountains are Engelmann spruce, shown here along with quaking aspen, and subalpine fir.

Subalpine forests

These are transition zones between alpine tundra and montane forests. They are defined by one or two dominant tree species, starting where they compose the majority of the canopy and ending at the treeline. As in any transition zone, they share some species with each adjacent habitat.

The subalpine forest of the Pacific Northwest coastal forests is dominated by mountain hemlock. Engelmann spruce and subalpine fir are the indicator species for subalpine forests in the Rocky Mountains, and those of the Sierra Nevada, Klamanth, and some western Great Basin ranges contain a mixture of whitebark pine, lodgepole pine, mountain hemlock, and foxtail pine. On the higher Great Basin ranges, limber pine and bristlecone pine (some of which are considered to be the oldest living organisms on earth) characterize the subalpine zone. In the East, subalpine forests contain white spruce, black spruce, balsam fir, and paper birch; they are, in fact, nothing more than a southerly extension of the boreal forest along the spine of the Appalachian Mountains.

One feature common to many subalpine forests is the stunted, misshapen conifer trees at the treeline. Known as krummholz (German for "crooked wood"), they have been shaped by the relentless wind, into a natural version of a giant bonsai tree. Some, called "banner trees," exhibit flagging, in which all branches grow on the leeward side of the prevailing winds. Other trees are so severely beaten by the wind that they form tangled, shrubby mats rather than a main trunk. The tree then tends to reproduce by layering, or rooting the tips or undersides of its branches whenever they touch moist soil, thus increasing the size of the mat.

Krummholz occurs when the sun and the friction caused by moving air warm the needles and stimulate their stomata to open, resulting in desiccation. Deprived of their moisture and unable to replace it because of frozen roots, the tips of the affected branches die back. Larger trees with more water reserves stored in their trunks suffer less than smaller trees. Curiously, this phenomenon seems to be more prevalent in the East, suggesting that the boreal forest species do not fare as well in fierce, cold windy weather as do western species.

Below Subalpine forests in the Great Basin region host bristlecone pines, the oldest known living organisms on earth. The most ancient specimens exceed 4,600 years in age.

Some common plants of western subalpine forests

Trees and shrubs

mountain hemlock	Engelmann spruce
subalpine fir	whitebark pine
lodgepole pine	foxtail pine
Douglas fir	Jeffrey pine
Pacific silver fir	subalpine larch
western hemlock	western redcedar
limber pine	bristlecone pine
alpine prickly currant	dwarf huckleberry
black spruce	bearberry
quaking aspen	western juniper
western serviceberry	red mountain heather
white mountain heather	creeping Oregon grape
shrubby cinquefoil	common snowberry
Rocky Mountain maple	mountain snowberry

Wildflowers

beargrass	blue columbine
subalpine buttercup	pussypaws
pearly everlasting	red columbine
Lewis' monkeyflower	pipsissewa
queen's cup beadlily	fireweed
elephant head	glacier lily
avalanche lily	arrowhead groundsel
heartleaf arnica	mountain bluebell
orange agoseris	western monkshood
Parry primrose	subalpine Jacob's ladder
star gentian	rose crown
giant red paintbrush	tall chiming bells
balsamroot	mountain gentian
twinberry honeysuckle	thimbleberry
American bistort	ballhead sandwort
mountain death camas	silverleaf phacelia
yellow monkeyflower	brook saxifrage
pink pussytoes	globemallow
Rocky Mountain lily	fringed grass of parnassus

Left Subalpine forests, which provide more shelter than alpine tundra, share many species with the vegetation zones of lower elevations. One of these, found in the herb layer, is blue columbine, the state flower of Colorado.

BOTANY AND YOU

WILD GARDENING

One of the most practical and rewarding ways to study botany is to start your own wild garden. A wild garden is not necessarily ground abandoned to follow its own course. More often, it is a carefully orchestrated re-creation of one or more specific habitats indigenous to your geographic locality, using native species. A homeowner in Illinois could convert the meadow behind the house into a replica of the tall-grass prairie that once covered his or her homesite, while in southern Arizona, a water-guzzling lawn could be replaced with the flora of the Sonoran desert.

In addition to being a pleasant and therapeutic pastime, wild gardening has low maintenance requirements. To be sure, there is occasional pruning, thinning, and transplanting to be done, but this is quite negligible compared to the attention required by the typical manicured garden. Lawns are, by definition, an early successional stage and a monoculture (a plant community composed primarily of one species), totally alien in the natural world, and therefore requiring a large expenditure of energy to maintain it.

Consider the large percentage of "free" time a homeowner spends mowing, raking, thatching, fertilizing, watering, and spraying the lawn, as well as clipping hedges, pruning shrubs, and weeding the garden. Add to this the cost, both financial and environmental, of the lawn mower or garden tractor, gas, oil, equipment maintenance, pesticides, herbicides, fertilizer, and water. At a point when both time and money are in short supply, it becomes increasingly difficult to justify the extravagance of a lush, manicured landscape.

In addition to the savings in terms of time and money, natural landscapes have other benefits. "Environment" is the buzzword of the nineties, and most people want to do their part. The native species in a wild garden can survive on the normal local precipitation, thus helping to conserve water. They require no pesticides because they have evolved defenses against most pests that plague domestic plants; a natural landscape also attracts wildlife, some of which provide natural pest control free of charge! Herbicides are out because a natural landscape is less susceptible to invasion by "weeds" than is a conventional garden; furthermore, a botanist will welcome most additions, whether intentional or not, and those unwelcome are easily removed by hand. Finally, a natural low-maintenance landscape may increase the value of your home as well as your enjoyment of it.

Choosing a theme

Natural ecosystems each have their own components, so to achieve a low-maintenance, healthy habitat, decide which habitat to copy and choose the appropriate plants. A hodge-podge of trees, shrubs, wildflowers, and lower plants thrown

Left and **right** Wild gardening, a concept spawned with the environmental movement of the seventies, has been steadily winning converts ever since. It involves the use of wild plants to achieve an attractive, natural-looking, low-maintenance landscape. If possible, it is often desirable to incorporate a water source, such as a pond or stream, into your wild garden for the sake of variety and the benefit of wildlife.

together on a whim is not likely to form a viable community and, in fact, you may lose many of your species in the first couple of growing seasons.

Always choose a wild-garden theme that is common to your area. Attempting a desert landscape in New England would be as fruitless as planting a tall-grass prairie high in the Rockies. Spend time researching local habitats through reading and especially through observation. Write down what grows where, which plants are frequently found together, which are never found together and the plants' maximum size, appearance, and life cycles.

Elements of a wild garden

A wild garden should not just be a collection of plants. There are other elements, both living and non-living, that make up the natural ecosystem, and you should include them in your garden to break up the monotony of a mass of vegetation, create accents, and lend a touch of realism.

Rocks are a part of many natural habitats and their inert forms contrast effectively with the organic shapes of plants. Instead of posing an obstacle to your lawn mower, an exposed rock ledge or a boulder deposited by a long-ago glacier can become the centerpiece of your wild garden, with pockets of soil in cracks and depressions to host shallow-rooted plants. A stone wall of the type found throughout the northeast, though not natural, still makes an attractive feature and may also create a cool, moist, micro-habitat, especially on the north side, for mosses, liverworts and lichens to thrive; you may be surprised at the many different types that take up residence.

In a wooded area, standing dead trees, or snags, and fallen logs and branches are a natural part of the landscape, so resist the urge to clean them out of your garden. Not only do they look natural, but they host a myriad of fungi which help decompose them and enrich the soil, a natural cycle which you should encourage. Snags and fallen logs are also home to a surprising variety of wildlife. If too much debris accumulates, detracting from the garden's appearance, rent or buy a chipper and turn the excess into a mulch that can be spread over the garden to supply soil-building humus.

You are indeed fortunate if you have a stream or pond, for not only is the water an attractive feature, but the moist soil along its edge is ideal for a wetland habitat. A moist depression can be turned into a pseudo-marsh, fed perhaps by rainwater collected from the gutters on your roof. Use your imagination! Even if your land is dry, you may be able to construct a pond that looks natural, using a recirculating pump to prevent the water becoming stagnant. Many wild gardening books provide detailed instructions for building ponds.

Over-crowding your garden with large or showy species may diminish their appeal, and be detrimental to their health. Use ground covers, plants that spread by runners or rhizomes to form a fairly dense layer of low vegetation, or mosses that form solid cushions of greenery, between featured trees and shrubs. Flowering ground covers often have small blooms which, rather than dominating the scene, add just a tinge of background color. When planting a meadow, use a generous portion of native grass seed; the meadow then appears more authentic, and the grass supports weaker-stemmed wildflowers that would otherwise become scraggly and prostrate. In fact, many native grasses create attractive, subtle accents used on their own.

A well defined system of footpaths, though not a "natural" feature, provides access to various parts of the garden and minimizes traffic in sensitive areas. Natural materials such as pine-bark mulch or a medium-colored gravel, tend to blend in better than harshly colored or obviously man-made ones.

Paths that cross a body of water will require either a foot bridge (**right**) or stepping stones. Try to include all parts of the habitat you wish to recreate, such as rocks, stumps, and logs; rocks may provide anchors for lichens, and fallen logs and stumps will host many invertebrates and essential decomposers like fungi (**left**) and bacteria.

Install small footbridges for crossing streams and narrow boardwalks to provide access to marshy areas while protecting wetland plants. Benches and logs provide welcome rest areas for enjoying the surroundings.

Obtaining seeds and native plants

Your first impulse may be to transplant plants from the wild. This is often the quickest and least expensive way of populating your garden, but it is seldom the best. Ethically, you should adopt the general rule of leaving plants where you find them. Even if you choose judiciously over a large area, the plants are part of a functioning ecosystem, and their absence leaves a void in that habitat. The main problem facing nature today is the encroachment of civilization and the accompanying loss or fouling of habitat, so habitats left relatively intact must be protected, to benefit future generations. Collecting plants removes not only that organism from that particular habitat, but all future generations that might have arisen from it, the food they would have produced, and the nutrients they would have returned to the soil. In addition, popular or rare plants may face extinction in the wild through well meaning collectors. Finally, many species do not transplant well, and some, not at all. Collecting them often just results in their death.

One instance where collecting plants from the wild may be justified and even encouraged is in areas slated for development. These plants are doomed anyway; better that they find a reprieve in your garden than be lost to the bulldozer's blade. Before staging plant rescue, you must obtain the landowner's permission, but most are cooperative.

Another method of obtaining plants from the wild is to collect ripe seeds, usually in late summer or fall. This has less impact on the habitat, because most species routinely produce many times more seeds than will ever grow into mature plants. To offset any negative effects, take only a few seeds from each plant, and collect from only a few plants in each location.

Most seeds have some form of germination inhibitor, a chemical that keeps them from sprouting at the wrong time. If you are storing seeds inside, these inhibitors must be deactivated, by a period of freezing that simulates winter, or by a certain amount of moisture, simulating melting snow or seasonal rains, before the seeds will germinate. From a nursery or library, obtain as much information as possible about the relevant species, to increase your chances of success.

The most ethical method of obtaining wild plants and seeds is to purchase them from nurseries. Many now specialize in native plants and offer mail-order service. Make sure, however, that the plants are nursery-raised and not collected from the wild; otherwise, you are paying someone else to negate your own good intentions.

WHERE TO GO

TALLGRASS PRAIRIE

Illinois
Belmont Prairie Preserve, Downers Grove; Fermi National Accelerator Laboratory; Gensburg-Markham Prairie, Markham; Goose Lake Prairie Nature Preserve, Morris; Illinois Beach State Park, Zion; James Woodworth Prairie Preserve, Chicago
Indiana
Hoosier Prairie Nature Preserve, Griffith
Iowa
Caylor Prairie, Spirit Lake
Kansas
Konza Prairie, Manhattan; Flint Hills Tallgrass Prairie, Rosalia
Minnesota
Minnesota Valley National Wildlife Refuge, Black Dog Unit, Bloomington; Malmberg Prairie, Polk County; Anna Gronseth Prairie, Wilkin County; Wahpeton Prairie, Walnut Grove
Missouri
Osage Prairie, Vernon County
Nebraska
Pawnee Prairie, Pawnee County
South Dakota
Sioux Prairie & Altamont Prairie, South Dakota State College, Brookings
Wisconsin
Scuppernong Springs Nature Area, Eagle; University of Wisconsin Arboretum, Madison

MIXED GRASS PRAIRIE

Kansas
Quivera National Wildlife Refuge, Stafford; Kirwin National Wildlife Refuge, Kirwin
Nebraska
Fort Robinson State Park, Crawford; Valentine National Wildlife Refuge, Brownlee; Oglala National Grassland, Crawford
North Dakota
Theodore Roosevelt National Park, Medora; Lostwood National Wildlife Refuge, Lostwood; J. Clark Sayler National Wildlife Refuge, Upham; Little Missouri National Grassland, Grassy Butte
Oklahoma
Black Kettle National Grassland, Cheyenne; Great Salt Plains National Wildlife Refuge, Vining; Wichita Mountains National Wildlife Refuge, Indiahoma
South Dakota
Badlands National Park, Interior; Wind Cave National Park, Hot Springs; Buffalo Gap National Grassland, Edgemont; Fort Pierre National Grassland, Fort Pierre; Sand Lake National Wildlife Refuge, Hecla; Grand River National Grassland, Lodgepole
Wyoming
Thunder Basin National Grassland, Upton

SHORTGRASS PRAIRIE

Colorado
Comanche National Grassland, Andrix; Pawnee National Grassland, Buckingham
Kansas
Cimarron National Grassland, Rolla
Montana
Red Rock Lakes National Wildlife Refuge, Lakeview; Charles M. Russell National Wildlife Refuge, Fort Peck; U. L. Bend National Wildlife Refuge, Valentine; Custer Battlefield National Monument, Crow Agency; Medicine Lake National Wildlife Refuge, Dagmar
Nebraska
Oglala National Grassland, Crawford; Crescent Lake National Wildlife Refuge, Oshkosh
New Mexico
Kiowa National Grassland, Clayton; Bitter Lake National Wildlife Refuge, Roswell
Texas
Rita Blanca National Grassland, Dalhart; Muleshoe National Wildlife Refuge, Needmore; Lake McClellan National Grassland Park, Alanreed

INTERMOUNTAIN GRASSLANDS

California
Modoc National Forest, Alturas
Clear Lake National Wildlife Refuge, Newell
Colorado
Dinosaur National Monument, Dinosaur
Idaho
Craters of the Moon National Monument, Arco; Curlew National Grassland, Holbrook; Caribou National Forest, Soda Springs; Sawtooth National Forest, Oakley; Camas National Wildlife Refuge, Camas
Nevada
Great Basin National Park, Baker; Charles Sheldon Wildlife Refuge, Denio Junction; Black Rock Desert, Sulphur; Humboldt National Forest, McDermitt; Humboldt National Forest, Mountain City; Humboldt National Forest and Ruby Lake; National Wildlife Refuge, Ruby Valley; Humboldt National Forest, Ely; Toiyabe National Forest, Austin; Smoke Creek Desert, Sand Pass
Oregon
Hart Mountain National Antelope Range, Plush; Malheur National Wildlife Refuge, Narrows; Malheur National Forest, Canyon City; Crooked River National Grassland, Madras; Ochoco National Forest, Mitchell
Utah
Bryce Canyon National Park, Bryce Canyon; Zion National Park, Springdale; Glen Canyon National Recreation Area, Ticaboo; Dixie National Forest, Cedar City

BOREAL FOREST

Alaska
Denali National Park, McKinley Park
Alberta
Wood Buffalo National Park, Fort Fitzgerald
Maine
Acadia National Park, Bar Harbor; Baxter State Park, Millinocket; Moosehorn National Wildlife Refuge, Calais
Manitoba
Whiteshell Provincial Park, Westhawk Lake
Michigan
Isle Royale National Park, St. Ignace; Hiawatha National Forest, Eckerman; Ottawa National Forest, Bergland
Minnesota
Lake Itasca State Park, Lake Itasca; Boundary Waters Canoe Area; Voyageurs National Park, International Falls
New Hampshire
White Mountain National Forest
New York
Adirondack Forest Preserve
Ontario
Algonquin Provincial Park, Whitney
Quebec
La Verendrye Provincial Park, Mont-Laurier
Saskatchewan
Prince Albert National Park, Christopher Lake
Tennessee
Great Smoky Mountains National Park, Gatlinburg
Vermont
Green Mountain National Forest
Virginia
Shenandoah National Park, Front Royal

NORTHERN HARDWOOD FOREST

Maine
Acadia National Park, Bar Harbor; Baxter State Park, Millinocket; Moosehorn National Wildlife Refuge, Calais
Michigan
Manistee National Forest, Big Rapids
New Hampshire
White Mountains National Forest
New York
Adirondack Forest Preserve
North Carolina
Pisgah National Forest, Asheville
Ontario
Algonquin Provincial Park, Whitney
Pennsylvania
Allegheny National Forest, Warren; Hickory Run State Park, White Haven; Rickett's Glen State Park, Red Rock; World's End State Park, Forksville
Tennessee
Great Smoky Mountains National Park, Gatlinburg
Vermont
Green Mountains National Forest
Virginia
Shenandoah National Park, Front Royal; George Washington National Forest; Jefferson National Forest

West Virginia
Monongahela National Forest
Wisconsin
Chequamegon National Forest, Ashland

OAK-HICKORY FOREST

Arkansas
Hot Springs National Park, Hot Springs; Ozark National Forest
Georgia
Chattahoochee National Forest
Kentucky
Mammoth Cave National Park, Mammoth Cave; Daniel Boone National Forest
Missouri
Mark Twain National Forest
New Jersey
Stokes State Forest, Tuttle's Corner
New York
Bear Mountain State Park, Jones Point; Harriman State Park, Willow Grove
North Carolina
Pisgah National Forest, Ashville
Ohio
Wayne National Forest
Pennsylvania
Hawk Mountain Sanctuary, Drehersville; Delaware Water Gap National Recreation Area, East Stroudsberg; Gettysburg National Military Park, Gettysburg
Tennessee
Great Smoky Mountains National Park, Gatlinburg; Cherokee National Forest
Virginia
Shenandoah National Park, Front Royal; George Washington National Forest; Jefferson National Forest
West Virginia
Monongahela National Forest

SOUTHERN PINELANDS

Alabama
Conecuh National Forest, Andalusia; Talladega National Forest, Tuscaloosa
Florida
Osceola National Forest, Lake City; Ocala National Forest, Ocala; Appalachicola National Forest, Tallahassee
Georgia
Oconee National Forest, Monticello; Piedmont National Wildlife Refuge, Macon; Laura S. Walker State Park, Waycross; Cumberland Island National Seashore
Louisiana
Kisatchie National Forest, Alexandria; Tensas National Wildlife Refuge, Tallulah
Mississippi
DeSoto National Forest, Hattiesburg; Homochitto National Forest, Meadville; Noxubee National Wildlife Refuge, Starkville
New Jersey
Wharton State Forest, Hammonton
North Carolina
Croatan National Forest, Havelock
South Carolina
Francis Marion National Forest, Goose Creek; Carolina Sandhills National Wildlife Refuge, Hartsville
Texas
Big Thicket National Preserve, Beaumont; Sabine National Forest, San Augustine
Virginia
Chincoteague National Wildlife Refuge, Chincoteague

FLOODPLAIN FORESTS

Locations are too numerous to mention here, but major river valleys, national wildlife refuges, national parks and forests, state parks and forests, Nature Conservancy preserves, and Audubon Society sanctuaries all offer excellent opportunities to observe floodplain forests.

PONDEROSA PINE FORESTS

Arizona
Kaibob National Forest, Flagstaff
Colorado
Rocky Mountain National Park, Estes Park; Pike National Forest, Colorado Springs; Gunnison National Forest, Gunnison
Idaho
Nez Perce National Forest, Grangeville
Montana
Glacier National Park, West Glacier; Custer National Forest, Ashland
New Mexico
Sante Fe National Forest, Santa Fe
South Dakota
Custer State Park, Custer; Black Hills National Forest, Rapid City
Utah
Ashley National Forest, Vernal; Fishlake National Forest, Fishlake
Wyoming
Bighorn National Forest, Sheridan; Medicine Bow National Forest, Laramie

DOUGLAS FIR FOREST

Alberta
Banff National Park, Banff
Jasper National Park, Jasper
Colorado
Rocky Mountain National Park, Estes Park
Idaho
Boise National Forest, Boise
Nez Perce National Forest, Grangeville
Montana
Glacier National Park, West Glacier
Kootenai National Forest, Libby
Lolo National Forest, Missoula
Wyoming
Grand Teton National Park, Moose

LODGEPOLE PINE FORESTS

Alberta
Banff National Park, Banff; Jasper National Park, Jasper
British Columbia
Mount Revelstoke National Park, Revelstoke
Colorado
Rocky Mountain National Park, Estes Park; Routt National Forest, Steamboat Springs
Idaho
Salmon National Forest, Salmon; Sawtooth National Recreation Area, Stanley
Montana
Helena National Forest, Helena; Glacier National Park, West Glacier
Oregon
Ochoco National Forest, Prineville
Utah
Ashley National Forest, Vernal
Wyoming
Bighorn National Forest; Yellowstone National Park; Grand Teton National Park, Moose; Bridger-Teton National Forest, Jackson

ASPEN FORESTS

Alaska
Denali National Park, McKinley Park
Alberta
Banff National Park, Banff; Jasper National Park, Jasper
Arizona
Kaibob National Forest, Flagstaff
Colorado
Rocky Mountain National Park, Estes Park; White River National Forest, Glenwood Springs; Gunnison National Forest, Gunnison
Idaho
Caribou National Forest, Soda Springs
Montana
Glacier National Park, West Glacier; Lewis and Clark National Forest, White Sulphur Springs
New Mexico
Cibola National Forest, Socorro
Utah
Cache National Forest, Ogden; Uinta National Forest, Provo
Wyoming
Medicine Bow National Forest, Laramie; Grand Teton National Park, Moose

PACIFIC NORTHWEST COASTAL FORESTS

Alaska
Kenai Fjords National Park, Seward; Glacier Bay National Park, Gustavus; Tongass National Forest, Juneau
British Columbia
Pacific Rim National Park, Ucluelet; Garibaldi Provincial Park, Whistler
Oregon
Wilamette National Forest, Oakridge; Mount Hood National Forest, Portland; Siskiyou National Forest, Grant's Pass
Washington
Olympic National Park, Port Angeles; North Cascades National Park, Sedro Woolley; Mount Rainier National Park, Ashford

REDWOOD FORESTS

California
Redwood National Park, Crescent City; Humboldt Redwoods State Park, South Fork; Muir Woods National Monument, Mill Valley; Grizzly Creek Redwoods State Park, Bridgeville; Jedediah Smith Redwoods State Park, Crescent City; Prairie Creek Redwoods State Park, Orick

BOGS

Connecticut
Black Spruce Bog, Mohawk Forest, Goshen
Maine
Acadia National Park, Bar Harbor; Crystal Bog, Crystal; Appleton Bog, Appleton; Meddybemps Heath, Calais
Massachusetts
Black Pond Bog, Vinal Nature Preserve, Cohasset; Thoreau's Bog, Concord
Michigan
Isle Royale National Park, Houghton
Minnesota
Voyaguers National Park, International Falls
New Hampshire
Wapack National Wildlife Refuge, Greenfield
New Jersey
High Point State Park, Sussex; Wharton State Forest, Hammonton
New York
Moss Lake Bog, Houghton; McLean Bogs, McLean
Ontario
Algonquin Provincial Park, Whitney
Pennsylvania
Tannersville Cranberry Bog Preserve, Tannersville
Rhode Island
Diamond Bog, Woodville
Vermont
Peacham Bog, Groton State Forest, Marshfield; Snake Mountain Bog, Weybridge
West Virginia
Cranberry Glades, Monongahela National Forest

MARSHES

There are too many sites to list. For directories contact:

National Wildlife Refuges: National Wetlands Inventory, U.S. Fish & Wildlife Service, Department of the Interior, 1730 K St., N.W., Washington, D.C. 20240, (202) 653-8726.
Nature Conservancy Preserves: The Nature Conservancy, 1800 North Kent St., Arlington, VA 22209.

EVERGLADES

Florida
Everglades National Park, Homestead; Loxahatchee National Wildlife Refuge, Loxahatchee; Big Cypress National Preserve, Monroe

CYPRESS SWAMPS AND SOUTHERN HARDWOOD SWAMPS

Arkansas
White River National Wildlife Refuge, St. Charles
Delaware
Pocomoke River Swamp, Gumboro
Florida
Blackwater River State Forest, Munson; Appalachicoloa National Forest, Tallahassee; Corkscrew Swamp Sanctuary, Immokalee; Big Cypress National Preserve, Monroe
Georgia
Okefenokee National Wildlife Refuge, Edith
Louisiana
Atchafalaya Swamp, Donner
North Carolina
Alligator River National Wildlife Refuge, East Lake; Merchants Mill Pond State Park, Sunbury
South Carolina
Four Holes Swamp, Harleyville; Congaree Swamp National Monument, Gadsden
Tennessee
Reelfoot National Wildlife Refuge, Walnut Log
Texas
Big Thicket National Preserve, Evadale
Virginia
Great Dismal Swamp National Wildlife Refuge, Suffolk

SAND DUNES

Florida
Canaveral National Seashore, Titusville; Gulf Islands National Seashore, Pensacola
Georgia
Cumberland Island National Seashore, St. Marys
Louisiana
Delta National Wildlife Refuge, Pilottown
Maryland
Assateague Island National Seashore, Berlin
Massachusetts
Cape Cod National Seashore, South Wellfleet; Parker River National Wildlife Refuge, Newburyport
New Jersey
Gateway National Recreation Area, Highlands; Island Beach State Park, Seaside Heights
New York
Fire Island National Seashore, Oak Beach
North Carolina
Cape Hatteras National Seashore, Manteo; Cape Lookout National Seashore, Beaufort
Oregon
Oregon Dunes National Recreation Area, Reedsport
Rhode Island
Ninigret National Wildlife Refuge, Charlestown
South Carolina
Cape Romain National Wildlife Refuge, McClellanville
Texas
Padre Island National Seashore, Corpus Christi
Virginia
Chincoteague National Wildlife Refuge, Chincoteague

ROCKY SHORES

Alaska
Glacier Bay National Park, Gustavus
British Columbia
Pacific Rim National Park, Ucluelet
California
Point Reyes National Seashore, Point Reyes Station; King Range National Conservation Area, Redway; Redwood National Park, Crescent City; Channel Islands National Park, Ventura; Cabrillo National Monument, San Diego
Maine
Acadia National Park, Bar Harbor
New Brunswick
Fundy National Park, Alma
Nova Scotia
Cape Bretton Highlands National Park, Cheticamp
Oregon
Oregon Islands National Wildlife Refuge, Brookings; Three Arch Rock National Wildlife Refuge, Cape Meares
Washington
North Cascades National Park, Sedro Woolley; Olympic National Park, Port Angeles; Willapa National Wildlife Refuge, Naselle; San Juan National Wildlife Refuge, Shaw Island

SALT MARSHES – ATLANTIC COAST

Connecticut
Salt Meadow National Wildlife Refuge, Grove Beach
Delaware
Bombay Hook National Wildlife Refuge, Smyrna
Florida
Merritt Island National Wildlife Refuge, Titusville
Maine
Rachel Carson National Wildlife Refuge, Wells; Scarborough Marsh, Falmouth
Maryland
Assateague Island National Seashore, Berlin; Blackwater National Wildlife Refuge, Church Creek
Massachusetts
Cape Cod National Seashore, South Wellfleet; Parker River National Wildlife Refuge, Newburyport

New Brunswick
Kouchibouguac National Park, Kouchibouguac
New Jersey
Edwin B. Forsythe National Wildlife Refuge, Brigantine
North Carolina
Alligator River National Wildlife Refuge, East Lake; Cape Hatteras National Seashore, Manteo
Prince Edward Island
Prince Edward Island National Park, Covehead
South Carolina
Cape Romain National Wildlife Refuge, Awendaw; Santee Coastal Reserve, McLellanville
Virginia
Chincoteague National Wildlife Refuge, Chincoteague

SALT MARSHES – GULF COAST

Florida
St. Marks National Wildlife Refuge, St. Marks; Gulf Islands National Seashore, Pensacola
Louisiana
Sabine National Wildlife Refuge, Hackberry; Rockefeller National Wildlife Refuge, Grand Chenier
Texas
Aransas National Wildlife Refuge, Austwell

SALT MARSHES – PACIFIC COAST

California
San Francisco Bay National Wildlife Refuge, Newark; Elkhorn Slough Estuarine Sanctuary; Salinas River Wildlife Management Area, Salinas; San Pablo Bay National Wildlife Refuge, San Pablo
Oregon
South Slough Estuarine Sanctuary
Washington
Willapa National Wildlife Refuge, Naselle; Dungeness National Wildlife Refuge, Dungeness; Nisqualla National Wildlife Refuge, Olympia; Grays Harbor National Wildlife Refuge, Olympia

MANGROVES AND TROPICAL FOREST

Florida
Everglades National Park, Homestead; Biscayne National Park, Homestead; J.N. "Ding" Darling National Wildlife Refuge, Sanibel; Great White Heron National Wildlife Refuge, Big Pine Key; Chassahowitzka National Wildlife Refuge, Chassahowitzka; Merritt Island National Wildlife Refuge, Titusville

GREAT BASIN DESERT

California
Modoc National Forest, Altura
Colorado
Dinosaur National Monument, Dinosaur
Idaho
Camas National Wildlife Refuge, Hamer; Curlew National Grasslands, Holbrook; Minidoka National Wildlife Refuge, Minidoka; Sawtooth National Forest
Nevada
Black Rock Desert, Sulphur; Great Basin National Park, Baker; Charles Sheldon National Antelope Refuge, Denio Junction; Humboldt National Forest; Toiyabe National Forest
Oregon
Hart Mountain National Antelope Refuge, Plush; Malheur National Wildlife Refuge, Narrows; Ochoo National Forest, Seneca
Utah
Bryce Canyon National Park, Bryce Canyon; Zion National Park, Springdale; Dixie National Forest, Panguitch; Glen Canyon National Recreation Area, Ticaboo

MOHAVE DESERT

Arizona
Hualapai Mountain State Park, Kingman
California
Joshua Tree National Monument, Twentynine Palms; California Poppy Reserve, Lancaster; Death Valley National Monument, Death Valley
Nevada
Desert National Wildlife Refuge, Indian Springs; Pharanagat National Wildlife Refuge, Alamo; Lake Meade National Recreation Area, Boulder City

SONORAN DESERT

Arizona
Desert Foothills Drive, Phoenix; South Mountain Park, Phoenix; Picacho Peak State Park, Picacho; Saguaro National Monument, Tuscon; Arizona-Sonoran Desert Museum, Tuscon; Organ Pipe Cactus National Monument, Ajo; Boyce Thompson Southwestern Arboretum, Superior
California
Anza-Borrego Desert State Park, Borrego Springs; Coachella Valley Preserve, Indio

CHIHUAHUAN DESERT

New Mexico
Living Desert State Park, Carlsbad; Carlsbad Caverns National Park, Carlsbad; White Sands National Monument, Alamagordo
Texas
Big Bend National Park

SUBALPINE FOREST AND ALPINE TUNDRA

California
Yosemite National Park; Kings Canyon National Park, Three Rivers; Sequoia National Park, Three Rivers; Klamath National Forest, Yreka
Colorado
Rocky Mountain National Park, Estes Park; Pike National Forest, Colorado Springs; Gunnison National Forest, Gunnison
Idaho
Salmon National Forest, Salmon
Maine
Mt. Katahdin, Baxter State Park, Millinocket; Cadillac Mountain, Acadia National Park, Bar Harbor
Montana
Glacier National Park, West Glacier; Bitterroot National Forest, Missoula
New Hampshire
The Presidential Range, esp. Mt. Washington, White Mountain National Forest
New York
Mt. Marcy, Lake Placid
Oregon
Mt. Hood National Forest, Cherryville
Utah
Uinta National Forest, Provo
Vermont
Mt. Mansfield, Mt. Mansfield State Forest, Stowe
Washington
Camel's Hump State Park, Ashford; Olympic National Park, Port Angeles; North Cascades National Park, Sedro Woolley
Wyoming
Grand Teton National Park, Moose; Yellowstone National Park

GLOSSARY

Abscission: the shedding of leaves, flowers, or fruits through the formation of an abscission zone at the point of attachment to the plant.
Adventitious roots: roots that develop along an above-ground stem or on leaves.
Aerobic respiration: the cellular breakdown of sugar or other food requiring oxygen and resulting in the release of energy.
Alternate: leaf arrangement of one leaf per node.
Anaerobic respiration: the cellular breakdown of sugar or other food that does not require oxygen but results in the release of energy.
Annual: a plant needing only one growing season to complete its life cycle.
Annulus: a membranous ring around the stem of a mushroom.
Anther: the pollen-bearing portion of a stamen.
Aperture: the adjustable opening of a camera lens which determines the amount of light passing through in a given time.
Apical meristem: the region of actively growing cells at the tip of a root or shoot.
Auxins: growth-regulating substances produced by plants.
Biennial: a plant requiring two growing seasons to complete its life cycle.
Bilateral symmetry: having two identical halves; symmetrical about one plane only.
Blade: the broad, flattened portion of a leaf.
Bulb: food-storage organ consisting of fleshy, modified leaves or leaf bases packed tightly around a small stem.
Calyx: the outer, protective part of a flower, made of sepals fused together to form a funnel, bowl or trumpet-shaped structure.
Chlorophyll: the green pigment present in the chloroplasts of plants and responsible for photosynthesis.
Chloroplasts: small granules within plants' cells which contain chlorophyll and other compounds active in photosynthesis.
Climax community: the stable, self-perpetuating community of vegetation at the culmination of ecological succession.
Compound: describing a leaf, flower or fruit divided into smaller, separate but similar units.
Corolla: inner protective part of a flower, composed of completely or partly fused petals.
Corm: an underground food-storage organ composed of a thickened stem usually covered by papery skin.
Cotyledon: an embryonic leaf within a seed, and the first to appear on germination.
Cross-pollination: the transfer of pollen from one plant to another of the same species.
Deciduous: a plant, especially a tree or shrub, which sheds all its leaves annually, at the end of the growing season.
Dicot: a class of flowering plants whose seeds have two cotyledons.
Desiccant: a drying agent.
Digestion: the enzyme-controlled conversion of complex, insoluble food molecules into simple, soluble molecules.
Entire: describing leaves or petals with smooth margins.
Fibrous roots: a spreading root system continuously branching into fine, hair-like structures.
Filament: the stalk of a stamen, bearing two anther lobes.
Frond: the leaf-like, spore-bearing organ of ferns and some other non-flowering plants.
Herbarium: a systematically arranged collection of dried, pressed, mounted, and labeled plant specimens.
Humus: the organic component of soil derived from decomposing plant or animal matter.
Hydrosere: the primary succession in a wet environment.
Hyphae: tubular, thread-like filaments of fungi.
Lanceolate: describing lance-shaped leaves, broad at the base and tapering to a point.
Lateral bud: a bud formed between a twig and the petiole of a leaf.
Leaflet: the sub-unit of a compound leaf.
Leaf scar: the scar left on a twig after the leaf is shed through abscission.
Lignin: the strengthening or stiffening substance that impregnates the cell walls of plants to varying degrees.
Linear: describing long, narrow leaves, bracts or petals, having parallel margins.
Margin: the edge of a leaf.
Midrib: the central vein of a pinnately-veined leaf from which lateral veins branch.
Monocot: the class of flowering plants whose seeds have one cotyledon.
Mycelium: the collective term for the hyphae of a fungus.
Node: the stem joint from which buds, side shoots and leaves form.
Opposite: describing a leaf arrangement in which two leaves are attached to the stem at the same point.
Ovary: the enlarged base of a pistil containing the ovules, which ultimately develop into seeds.
Ovate: describing egg-shaped leaves, bracts or petals, which are broadest at the base.
Ovules: immature seeds in the ovary containing the eggs.
Palmate: describing hand-shaped leaves, with deep lobes radiating from a single point.

Peduncle: the stalk of a flower or flower cluster.
Pedicel: the stalk of an individual flower within a flower cluster.
Perennial: a plant, usually herbaceous, whose life cycle exceeds two years.
Perianth: the collective term for the sepals and petals of a flower, in which the two structures are indistinguishable from one another.
Petal: the basic, flattened unit of the corolla, often brightly colored.
Petiole: the stalk of a leaf.
Phloem: tube-like conductive tissue which transports food up or down in vascular plants.
Pinnae: the primary sub-units of a deeply divided leaf or frond.
Pinnate: describing opposite leaflets or veins branching from both sides of a common axis.
Pioneers: the first plants in any successional stage to become established.
Pistil: the central, female organ of a flower consisting of a stigma, a style, and an ovary.
Pollen: the male reproductive cells of a plant, formed in the anthers.
Pollen tube: a structure that develops from a pollen grain and conducts the male cells from the stigma through the style to an ovule.
Pollination: the transfer of pollen from an anther to a stigma.
Primary succession: the ecological succession beginning in an area devoid of soil.
Prothallus: the initial stage of a fern's life cycle.
Protoplasm: the living, translucent liquid substance of a cell including cytoplasm and the nucleus.
Rachis: the central axis of a compound leaf or raceme or spike on which flowers or leaflets are carried.
Radial symmetry: having identical halves divisible by an infinite number of planes.
Receptacle: the swollen tip of a stem to which all of the flower parts are attached.
Rhizoids: tiny, root-like structures of certain non-vascular plants that function as anchors.
Rhizome: a horizontal, underground stem acting as a food storage organ.
Respiration: the cellular breakdown of sugars and other foods resulting in the release of energy.
Secondary succession: ecological succession in an area where soil has already formed.
Sepal: the primary leaf- or petal-like unit of the calyx whose major function is to protect the unopened flower bud.
Seral stage: one stage in the process of ecological succession.
Sere: the entire process of ecological succession.
Serrate: describing leaves with forward pointing, saw-like teeth on the margins.
Simple leaf: the opposite of compound; a broad, undivided leaf.
Sori: the spore-bearing structures of ferns.
Stamen: the male reproductive organ of a flower consisting of a filament and an anther.
Stigma: the sticky tip of the pistil which receives the pollen grains.
Stolon: a prostrate, creeping stem growing along the surface of the ground, and rooting from the nodes, producing fuller stems or plantlets.
Stoma: a minute pore, usually on the underside of a leaf, which opens and closes to regulate transpiration and gas exchange.
Strobili: the catkin-like, spore-bearing flower spikes of conifers.
Sucker: a shoot arising from below ground, usually from the roots and remote from the main stem.
Style: that portion of the pistil connecting the stigma with the ovary in a female flower.
Taproot: the main anchoring root of a plant, especially a tree, often reaching deep into the ground.
Tendril: a slender, modified leaf or stem that coils around a support on contact and aids climbing.
Terminal bud: the topmost bud at the end of a twig.
Transpiration: the loss of water vapor, through stomata, from leaf and stem surfaces.
Tuber: a swollen, starchy underground stem or thickened, fleshy root functioning as a food storage organ, and a means of surviving drought or cold.
Undulate: describing a leaf, sepal or petal with a wavy or crimped margin.
Veins: strands of tissue which support a leaf and consist of xylem and phloem, the internal conducting elements of a leaf.
Weathering: the chemical or physical breakdown of rock into its smallest units, which form the inorganic portion of soil.
Whorl: describing three or more leaves attached at a single point on the stem, like the spokes of a wheel.
Xerosere: ecological succession occurring in a relatively dry environment.
Xylem: tube-like vascular tissue that forms the wood, and conducts water and dissolved substances upward in a plant.

INDEX

page numbers in **bold** refer to illustrations

CONTACTS

NATIONAL BOTANICAL ORGANIZATIONS

Environmental Defense Fund, 1616 P St., NW Suite 150, Washington, D.C. 20009.
The Nature Conservancy, 1800 N. Kent St., Suite 800, Arlington, VA 22209.
American Horticultural Society, P.O. Box 0105, Mt. Vernon, VA 22301.
The Garden Club of America, 598 Madison Ave, New York City, NY 10461.
National Arbor Day Foundation, 100 Arbor Ave, Nebraska City, NE 68410.
Canadian Wildflower Society, 35 Bauer Crescent, Unionville, Ontario L3R 4H3.
Royal Botanical Gardens Members Association, Box 339, Hamilton, Ontario L8N 3H8.
Canadian Botanical Association, Department of Botany, University of British Columbia, Vancouver, British Columbia V6T 2B1.
National Council of State Garden Clubs, Inc., Operation Wildflower, P.O. Box 860, Pocasset, MA 02559.
National Council of State Garden Clubs, Inc., 4401 Magnolia Ave, St. Louis, MO 63110.

SOURCES OF NATIVE PLANTS AND SEEDS

Baldwin Seed Company of Alaska, Box 3127, Kenai, AK 99611.
Bill Dodd's Rare Plants, P.O. Box Drawer 377, Semmes, AL 36575.
Alberta Nurseries & Seeds, Ltd, Box 20, Bowden, Alberta T0M 0K0.
Beaverlodge Nursery, Ltd, Box 127, Beaverlodge, Alberta T0H 0C0.
Izard Ozark Natives, P.O. Box 32, Berryville, AR 72616.
Southwestern Native Seeds, Box 50503, Tucson, AZ 85703.
Alpenglow Gardens, 13328 King George Highway, North Surrey, British Columbia V3T 2T6.
British Columbia Nursery Trades Association, Suite #101-A, 15290-103A Ave, Surrey, British Columbia V3R 7A2.
The Plant Man Yes, 1017 2nd St, Santa Rosa, CA95405.
Wildwood Farm, 10300 Sonoma Hwy. Kenwood, CA 95452.
Dean Swift Seed Co, P.O. Box B, Jaroso, CO 81138.
Salter Tree Farm, Rt. 2, Box 1332, Madison, FL 32340.
Wildflowers From Nature's Way, R.R. #1, Box 62, Woodburn, 1A 50257.
Northplan Seed Producers, P.O. Box 9107, Moscow, 1D 83843.
Windrift Prairie Shop and Nursery, R.D. #2, Oregon, IL 61061.
Sharp Bros. Seed Co, P.O. Box 140, Healy, KS 67850.
Louisiana Nature and Science Center, 11000 Lake Forest Blvd, New Orleans, LA 70127.
Manitoba Nursery & Landscape Association, 104 Parkside Dr, Winnipeg, Manitoba R3J 3P8.
Environmental Concern Inc, 210 West Chew Ave, P.O. Box P, St. Michaels, MD 21663.
Far North Gardens, 16785 Harrison, Livonia, MI 48154.
Prairie Restorations, Inc, P.O. Box 327. Princeton, MN 55371.
Missouri Wildflowers Nursery, Rt. 2, Box 373, Jefferson City, MO65101.
Lawyer Nursery, Inc, 950 Hwy. 200 West, Plains, MT 59859.
Niche Gardens, Rt. 1, Box 290, Chapel Hill, NC 27514.
Stock Seed Farms, Inc, R.R. #1, Box 112, Murdock, NE 68407.
Plants of the Southwest, 1812 2nd St, Santa Fe. NM 87501.
Atlantic Provinces Nursery Trades Association, 130 Bluewater Rd, Bedford, Nova Scotia B4A 1G7.
Downham Nursery, Inc, 626 Victoria St, Strathroy, Ontario N7G 3C1.
Woodland Nurseries, 2151 Camilla Rd, Mississauga, Ontario L5A 2K1.
Sheridan Nurseries, 1116 Winston Churchill Blvd, Oakville, Ontario L6J 4Z2.
Canadian Nursery Trades Association, 1293 Matheson Blvd. Mississauga, Ontario L4W 1R1.
Northwest Biological Enterprises, 23351 SW Basky Dell Lane, West Linn, OR 97068.
Appalachian Wildflower Nursery, Rt. 1, Box 275A, Reedsville, PA 17084.
Wildflower Patch, 442 RC Brookside Dr., Walnutport, PA 18088.
Greenhedges, 650 Montee de Liesse, Montreal, Quebec H4T 1N8.
Saskatchewan Nursery Trades Association, Box 460, Carnduff, Saskcatchewan S0C 0S0.
Oak Hill Farm, 240 Pressley St, Clover, SC 29710.
Woodlanders, Inc., 1128 Colleton Ave, Aiken, SC 29801.
Echinational Plant Products, 602 Jefferson St, Vermillion, SD 57069.
Native Gardens, Rt. 1, Box 494, Fisher Lane, Greenback, TN 37742.
Sunlight Gardens, Inc, Rt. 3, Box 286TX, London, TN 37774.
Green Horizons, 218 Quinlan, Suite 571, Kerrville, TX 78028.
Gunsight Mountain Ranch & Nursery, P.O. Box 86, Tarpley, TX 78883.
Bambert Seed Co, Rt. 3, Box 1120, Muleshoe, TX 79347.
Mid-Atlantic Wildflowers, S/R Box 226, Gloucester Point, VA 23062.
Frosty Hollow Nursery, Box 53, Langely, WA 98260.
Little Valley Farm, R.R. 1, Box 287, Richland Center, WI 53581.

ACKNOWLEDGMENTS

Quarto would like to thank Rick Imes and Photo/Nats for providing photographs and for permission to reproduce copyright material. With the exception of Rick Imes and Moira Clinch, all the photographers credited are represented by Photo/Nats.

Key: bl = bottom left; br = bottom right; tr = top right; tl = top left; l = left; r = right; t = top; c = center; b = bottom

Liz Ball 74
Jean Baxter 126bl
Philip Beaurline 49bl; 49br; 81b; 84t; 84b; 84c; 107; 133; 134t; 137cr
Louis Borie 221; 32b; 72; 95t
Paul M. Brown 26c; 78; 116tl; 137cl; 139r; 144
Gay Bumgarner 77b; 148
Al. Bussewitz 125b
Priscilla Connell 77t; 791; 121t; 139br
Greg Crisci 44
Deborah Crowell 62
Betsy Fuchs 110
Michael Goodman 90t; 104
Carl Hanninen 6; 23t; 64t; 73; 92t
Edward S. Hodgson 18; 139bl
Valorie Hodgson 32t; 41c; 41b; 47t; 50tl; 50tr; 51tl; 51tr; 71t; 123; 130; 131tr
Hal Horwitz 26t; 93t; 99; 115t; 115b; 129tl; 131br
Don Johnston Photos 20t; 28; 47bl; 47br; 48; 63;68b; 94t; 102b; 106; 111; 112t; 112b; 113; 143
Sydney Karp 29; 136
Dorothy Long 87
John A. Lynch 23b; 24br; 24t; 26b; 30b; 37; 60t; 61b; 66t; 86; 116b; 127r; 151
Robert E. Lyons 17r; 25; 49t; 79r;
Stephen G. Maka 19; 34; 118; 119r; 120
Jeff March 24bl; 51b; 68t; 83t; 105; 147l; 147r
A. Peter Margosian 42; 138t; 138b; 140
Iuan Massar 53
Mary A. Mather 8br; 57; 90b
Herbert B. Parsons 8tr; 121b
Ann Reilly 54; 141; 149
John J. Smith 13; 14; 171; 59c; 59b; 601; 92bl; 92br; 94b; 100; 108; 117; 125t; 126t; 126bc; 128; 129b; 132; 134b; 137t
David M. Stone 27; 36; 38; 41t; 56; 67c; 67b; 71b; 81t; 95bl; 97t; 98r; 109; 116tr; 119l; 124; 129tr; 137b
Eugene H. Walker 65b; 70
Mrs. Eugene H. Walker 52; 80b; 142; 145t; 145b
Muriel V. Williams 31; 35t; 58
Marilyn Wood 11; 22r; 96; 122; 135

Moira Clinch 3
Rick Imes 20b; 21; 30t; 59t; 64b; 80t; 82b; 91; 93b; 95br; 97b; 98l; 101t; 101b; 102t; 103; 146; 150